EASTERN INFLUENCES ON NEUROPSYCHOTHERAPY

EASTERN INFLUENCES ON NEUROPSYCHOTHERAPY

Accepting, Soothing, and Stilling Cluttered and Critical Minds

edited by

Giles Yeates and Gavin Farrell

Volume 1 in the Specialist Topics of the Neuro-Disability and Psychotherapy Series

Routledge
Taylor & Francis Group

LONDON AND NEW YORK

First published 2018
by Routledge
2 Park Square, Milton Park, Abingdon, Oxon OX14 4RN

and by Routledge
711 Third Avenue, New York, NY 10017

Routledge is an imprint of the Taylor & Francis Group, an informa business

British Library Cataloguing-in-Publication Data
A catalogue record for this book is available from the British Library

Library of Congress Cataloging-in-Publication Data
A catalog record has been requested for this book

ISBN: 9781782206156 (pbk)

Typeset in Palatino
by The Studio Publishing Services Ltd
www.publishingservicesuk.co.uk
email: studio@publishingservicesuk.co.uk

CONTENTS

ABOUT THE EDITORS AND CONTRIBUTORS

Editors

Dr Gavin Farrell is co-editor of the journal *Neuro-Disability & Psychotherapy*. He works as a Consultant Clinical Neuropsychologist in neuro-rehabilitation. He is trained in Intensive Short-Term Dynamic Psychotherapy (ISTDP). Gavin's specialist area of interest is in working with medically unexplained neurological symptoms such as psychogenic non-epileptic seizures. Gavin also is a practitioner of Daoist Qi Gong, Tai Ji and Kung Fu, and has an interest in traditional Chinese medicine.

Dr Giles Yeates is editor of the journal and book series *Neuro-Disability & Psychotherapy*, in addition to the *Brain Injury* book series. As a clinical neuropsychologist in community neuro-rehabilitation, his clinical work and research focuses on the innovation of psychological therapies and support of relationships following acquired brain injury. Giles is also a longstanding practitioner and teacher of Tai Ji and Kung Fu, and studies under the Daoist Monks in Wudang Mountain, China. Giles is currently innovating the use of Tai Ji in neuro-rehabilitation.

Contributors

Fiona Ashworth is a senior lecturer at Anglia Ruskin University and honorary clinical psychologist at the Evelyn Community Head Injury Service, Cambridge Community Service. Her area of interest is in the role of compassion in mental health and well-being, specifically in unravelling the emotional and cognitive consequences of acquired brain injury and interventions that might alleviate these problems. She aims to make more sense of the emotional experiences (such as shame) and psychological responses to brain injury (such as self-criticism) and whether interventions with a compassion focus are likely to be a useful tool to alleviate these difficulties. Fiona is particular interested in the application and utility of Professor Gilbert's compassion focused therapy with people with acquired brain injuries and their relatives. Her interest in the role of compassion in well-being extends beyond brain injury and into other mental health groups as well as non-clinical groups (e.g., the role of compassion in adolescents with potential mental health problems). Her interests not only span psychosocial aspects but also biological and neural substrates (e.g., hormonal responses to stress in people with acquired brain injury).

Neil Carrigan is a Clinical Psychologist in the Psychological Therapies Service at Avon & Wiltshire Mental Health Partnership NHS Trust. He obtained a PhD in Psychology from the University of Leeds and a Doctorate in Clinical Psychology from the University of Bath. He is also accredited as a Cognitive Behavioural Therapist with the British Association of Behavioural & Cognitive Psychotherapies. He has research interest in trans-diagnostic approaches to mental health difficulties and the use of mindfulness based therapies.

Niels Detert is a clinical neuropsychologist at the John Radcliffe Hospital in Oxford, UK, working with acute and out-patient neurology and neurosurgery patients. He has a special interest in using mindfulness and experiential dynamic psychotherapy to help people with neurological and functional/dissociative disorders to cope better, reduce anxiety, depression and symptoms, and regain a full appreciation of life.

Laura Douglass is an honorary assistant psychologist at the John Radcliffe Hospital in Oxford, UK. She is an undergraduate at Cardiff University on a third year clinical placement focussing on clinical neuropsychology and mindfulness research.

Leon Dysch is a Clinical Neuropsychologist at St Martin's Hospital in Bath, working in the Community Neuro and Stroke Service and Neuropsychology Outpatient Service. He holds an Honorary Senior Lecturer position at the University of Bath where he is involved in convening and contributing to the Neuropsychology teaching for the Doctorate in Clinical Psychology. He has a special interest in the psychological impact and management of long-term neurological conditions and, in particular, the application of third wave therapies within this context.

David Gillanders is a senior lecturer in clinical psychology at the University of Edinburgh and academic director of the university's clinical psychology training programme. An experienced ACT clinician and trainer, he has an interest in the application of ACT to improving outcomes in chronic disease populations.

Sarah Gillanders is a chartered clinical psychologist and consultant clinical neuropsychologist working in private practice for Case Management Services in Edinburgh. She has specialised in neuropsychology since 2006 and has special interests in multiple sclerosis and the psychological adjustment to neurological illness and brain injury. She currently carries out neuropsychological assessments and therapy in the clinical and medicolegal settings.

Anke Karl is a Senior Lecturer in Clinical Psychology at the University of Exeter with great interest in translational research. Her major research interest are biopsychological mechanisms of posttraumatic stress disorder and its successful treatment. She uses psychophysiological and neuroimaging techniques and her current research focuses on protective mechanisms that facilitate adaptive emotion regulation for recovery from psychological trauma. Anke is also a trained CBT practitioner and specialised in traumafocused therapy for PTSD.

David McLaughlin is a trainee clinical psychologist at Plymouth University. His research focus is on interventions to promote reflective capacity.

Dr Rebecca Poz is a clinical psychologist and a clinical neuropsychologist chartered with the British Psychological Society. She has worked in the specialism of older people within the NHS since 2001, and currently stands as the lead advisor for older people at the BPS Division of Neuropsychology Policy Unit. Rebecca also works in private practice with a range of conditions, but predominantly acquired brain injuries and post-traumatic conditions.

Tamara A. Russell is a clinical psychologist, neuroscientist, martial artist and mindfulness trainer. She works nationally and internationally delivering teaching and training on the topic of mindfulness in health and education predominantly working with her Body In Mind Training methodology. In her private practice she specialises in working with individuals with severe and complex issues including psychosis and bipolar.

Dr Aneesh Shravat is a Counselling Psychologist chartered with the British Psychological Society. He currently works within the NHS for the Oliver Zangwill Centre providing specialist rehabilitation after brain injury. His research interests include working with identity and the role of the therapeutic relationship within holistic rehabilitation.

Tiago P. Tatton-Ramos, psychologist, master in Religious Studies (UFJF-MG-Brazil), PhD in Psychology (UFRGS-RS-Brazil), postdoc researcher in Psychiatry and Behavioral Sciences (UFRGS-RS-Brazil) and co-founder of INICIATIVA MINDFULNESS BRAZIL (www.iniciativamindfulness.com.br).

Oliver J. Tooze is a clinical psychologist at Fettle House, Bodmin Hospital. His main area of work is with people who have long-standing mental health difficulties and who may experience psychosis. He has a special interest in contextual behaviour approaches, in particular their use in chronic conditions.

The 2018 specialist topics in neuro-disability and psychotherapy: eastern influences on neuropsychotherapy

It is a great pleasure to introduce the first of our specialist topics collected works within neuro-disability and psychotherapy. Following three years of a traditional journal format, *Neuro-Disability & Psychotherapy* enters a new phase of publication as an annual book series of collected works on sub-topics within the field. The original journal hosted both diverse stand-alone publications and, in its second year in 2014, experimented with a double themed issue on the influences of eastern philosophical and spiritual ideas on innovations within neuropsychotherapy. This format was something we have been keen to repeat, hence we are particularly excited by the new direction of the journal going forward.

The complexity and nuances within the endeavour of neuro-psychotherapy are marked. Neuro-disability is characterised on a case by case basis by unique constellations of varying physical and cognitive difficulties, altered emotional experience, distinct relational patterns, and diverse social contexts. This heterogeneity necessarily spawns the evolution of multiple pockets of specialist sub-topics and connecting conversations, be it a community within a particular therapy modality sharing their perspective on applying the approach with developmental, acquired, or progressive conditions, or several

therapists approaching one form of psychological distress in a neurological condition from diverse theoretical orientations. We are excited
to present a diverse spectrum of these specialist conversations to the
readership over the forthcoming years, disseminating the innovative
and varied work of clinicians and researchers.

Topics earmarked for future publications within this series include
psychotherapy and aphasia; post-traumatic stress in neurological conditions;
and *brief psychotherapy interventions with neurological conditions.* Each of
these future works will be co-edited by one of us and a guest co-editor,
a leader in their specialist field. It will be an honour for us to be joined
each year by inspiring practitioners and communicators within neuro-
disability and psychotherapy sub-specialisms.

We kick-start the new series in 2018 by a republishing of existing
articles from *Neuro-Disability & Psychotherapy,* re-assimilated in a
novel combination for the *Specialist Topics* series. We revisit the 2014
double-themed issue on eastern approaches to neuro-rehabilitation.
We include the entire contents of the 2014 double issue plus two additional articles from subsequent journal issues that fall within the same
topic. First, we include Yeates' theoretical article on the Daoist/Positive
Psychology concept of Flow State Experience as a guiding principle
for the application of tai ji for people with neurological conditions
(originally published in the first 2015 issue). Second, we incorporate
an article by Carrigan and colleagues on Acceptance and Commitment
Therapy (ACT) with multiple sclerosis, to add to the other ACT articles in the original double issue.

We arrive at a fine collected works as a result. As we discussed in
our original editorial for the double issue (included in this collection),
the applied potential of ideas from Buddhist, Daoist, and Vedic spiritual, psychological, and mind-body practices is significant. All share a
common focus on altering the individual's relationship to their mental
contents, yet attempt to do so by very varied practices. Both of these
characteristics are hugely important for neuropsychotherapy. Neuro-
disability in all of its forms commonly makes it harder for people to
use higher forms of attentional control to manage their mental experience, be it to access alternative thoughts in the face of negative cognitions or reflect on repeating themes from interpreted unconscious
material. The aforementioned heterogeneity in needs inevitably means
that not every user of neurological services will optimally benefit from
a traditional seated talking therapy approach. This may be a result of

cognitive and physical difficulties (e.g., a short attention span or low arousal state precluding seated verbal exchange as a meaningful vehicle for psychological change) and/or the barrier of a service user's reluctance to begin something called psychotherapy to contemplate adjustment when they would rather have, say, more physiotherapy of persisting hemiplegia following their stroke. Both these challenges require innovative approaches to psychological support that deviate from traditional psychotherapy formats and perhaps rise to offer multiple positive gains, such as both psychological and physical functioning benefits (as in the case of yoga and tai ji).

We hope you, the readers, enjoy our first instalment of *Neuro-Disability & Psychotherapy: Specialist Topics;* we look forward to presenting more focused conversations and disseminated works in the years to come. We anticipate that the new and exciting ideas and directions for practice contained in this new series will stimulate innovations in clinical practice around the world in neurological services. Finally, if a reader is connected within a network of clinicians and researchers focused on a specialist topic within neuropsychotherapy, and would like to join one of us to guest co-edit a future volume of this series, do get in touch. Your ideas will be warmly welcome.

2014 Special Issue*

Giles Yeates and Gavin Farrell

P roviding emotional support for people with neurological conditions is confronted by a paradox: in addressing thoughts and
feelings that feel distressing and out of control many traditional
talking therapies intrinsically demand mental control, the very thing
many clients complain they do not have. Examples would be the need
to be able to intentionally access alternative thoughts during cognitive
restructuring in cognitive–behavioural approaches; abstract from a
behavioural experiment in the same modality to inform reappraisal;
initiate a process of re-evaluation or introspection; make new connections, or at a basic level expect a significant level of change in lived
self-experience through predominantly verbal-based interventions. As
clinicians in neurological services we are all too aware of the myriad
diversity in constellations of physical, cognitive, emotional, and social
problems across service users and how they will differentially challenge the traditional models of psychotherapy.

This book, the first of the series dedicated to a particular topic,
presents a range of approaches that step out of the traditional Western

---

* Originally published as the Editor's Column in 2014 in *Neuro-Disbility & Psychotherapy*,
2(1/2): vii–viii

psychotherapy mould, to offer new paradigms to address the afore-mentioned diversity in service user needs. They all share a direct or indirect heritage from Eastern spiritual practices, with their common focus to change the relationship of the individual to their mental contents, without necessarily aiming to change the mental contents themselves. They do this in different ways, with the value of accep-tance and commitment therapy (ACT) for diverse aspects of multiple sclerosis explored by Oliver Tooze and colleagues, together with Sarah Gillanders and David Gillanders. Compassion focused therapy (CFT) is then discussed in relation to acquired brain injury (Fiona Ashworth) and dementia (Rebecca Poz), the applicability of the model is clear in both acquired and progressive conditions. While the focus of these models may be on changing internal relationships between aspects of self-experience, the role of the body as an additional vehi-cle for managing mental contents is taken up in the next three chap-ters. A fusion of CFT and yoga is described by Aneesh Sharavat, the potential application of an embodied mindfulness intervention based on martial arts movements is introduced by Tamara Russell. The issue is concluded by Detert and Douglass' analysis of data from a mind-fulness-based group intervention for diverse client groups in a busy neurological service, nicely dove-tailing the twin themes of this issue: compassionate observation of mental contents alongside bodily awareness, to derive new relationships to distressing aspects of subjective experience in neurological conditions.

This book contains many riches in terms of theory, concepts, and practices. Conceptual chapters, clinical case studies, and a group eval-uation are all represented, which we feel demonstrate the value of our pluralist editorial policy. We trust readers agree and find value in this diversity, perhaps finding at least one or two ideas to approach an unusual presentation from a fresh perspective.

CHAPTER ONE

An exploration of acceptance and commitment therapy for chronic pain in multiple sclerosis*

Oliver J. Tooze, Anke Karl, Leon Dysch, and David McLaughlin

Introduction

This chapter begins by looking at the impact of pain in multiple sclerosis (MS), current treatment options, and areas of need that are not fully addressed. The insufficiency of current medical approaches to managing pain in MS is discussed and the need for treatments that target the psychological consequences of MS is introduced in this context. The potential for therapies that target cognitive and behavioural responses to symptoms is raised and a novel psychological approach to treating chronic pain is explored. Specifically, the targeting of psychological flexibility and acceptance within the acceptance and commitment therapy (ACT) approach. ACT aims to improve people's ability to live with treatment resistant symptoms rather than focusing on symptom reduction and as such, has the potential for treatment benefits in multiple domains. The ACT approach is briefly outlined and discussed in terms of its usefulness beyond pain management for people with MS (pwMS). As there is limited evidence about the use of ACT specifically with chronic pain in MS, studies using convergent approaches with a variety of chronic

---

* Originally published in 2014 in *Neuro-Disability & Psychotherapy*, 2(1/2): 1–18.

and progressive conditions are reviewed including correlational studies of proposed treatment processes and outcome studies of clinical interventions. An illustration of aspects of treatment using ACT is provided in the form of a vignette. This chapter acknowledges the lack of high quality evidence specifically supporting the use of ACT for chronic pain in MS at present and draws on wider literature on ACT processes and treatment outcomes with other patient groups to build a theoretical argument for further research in this area.

Search methods for the identification of studies

A search was conducted across three databases: PsycINFO, Web of Knowledge, and PubMed using the terms "multiple sclerosis", "pain", "acceptance", "acceptance and commitment therapy", and "mindfulness". Reference lists from key papers were also examined to identify further sources. Articles were reviewed by title and abstract for potential relevance. Key author searches and citation searches for key papers were also completed.

Background

MS is a progressive neurological condition affecting around 110 people per 100,000 population in England and Wales (Richards et al., 2002). It is the most common disease of the central nervous system affecting young adults (Multiple Sclerosis Trust, 2008) and is associated with significant psychosocial impairment (Foley & Brandes, 2009). MS tends to affect young adults during their working lives (Sadovnik et al., 1992) meaning the costs resulting from MS related disability to the individual and society are considerable (Whetten-Goldstein, et al., 1998).

Nearly all MS patients experience sensory alterations including pain, numbness, and tingling at some point in their illness (Miller, 2001). Prevalence estimates for chronic pain range from 48% to 65% (Khan & Pallant, 2007; Moulin et al., 1988; O'Connor et al., 2008). For those with MS, pain has been demonstrated to have a detrimental impact on quality of life (Hawthorne et al., 1999), health related quality of life (Forbes et al., 2006), psychological well-being and independent living (Hawthorne et al., 1999) as well as being associated with poorer mental health (Stewart & Ware, 1992). Medication is frequently used for the treatment of pain in MS (Heckman-Stone & Stone, 2001), however, there is still no satisfactory medical approach to managing pain in MS (Rossi et al., 2009) and currently no robust support for the efficacy of any one approach (O'Connor et al., 2008). There are also perceived barriers to accessing pain treatment including lack of accessible pain or MS specialists, side effects of medication, fear of taking medication/dependence, lack of finances, and the belief that nothing works for pain (Khan & Pallant, 2007). Treatment for pain has also been found to account for nearly 30% of the total use of medications for the management of all MS related symptoms (Solaro & Uccelli, 2011). There are many different types of pain associated with MS, as well as a range of other potential impairments such as fatigue, spasticity, balance problems, and visual disturbances. There are also differences in MS type, most commonly including phases of exacerbation and remission. These factors, along with the unpredictable and progressive nature of the disease, present challenges for those living with the condition, and to the development of effective treatments.

Currently available treatments do not offer full symptom relief and do not address the significant psychosocial component of this condition. While further research into symptom reduction is critical, there is also a pressing need to investigate how best to support people living with this condition within the current limited treatment environment.

The evidence for cognitive behavioural approaches to managing the psychological consequences of MS is promising. Cognitive and behavioural responses to symptoms have been related to health related functional impairment and have been found to be more closely related to distress than illness severity (Dennison et al., 2009). Importantly, these responses are potentially modifiable. The available evidence suggests that cognitive behavioural therapy (CBT) is effective in the treatment of depression with this group of people and in helping people adjust to the condition. The Cochrane review (Thomas et al., 2006) called for further pragmatic research to establish whether CBT is effective for chronic pain with this group of people as well as consideration of the active processes.

ACT builds upon cognitive-behavioural principles and has demonstrated potential as a method of enabling people to live well with chronic pain. Data from chronic pain treatment programmes using ACT suggest that improved outcomes may not be dependent on changes in pain symptoms (McCracken & Gutierrez-Martinez, 2011; McCracken et al., 2005; Wicksell et al., 2010) but are associated with changes in acceptance and psychological flexibility—the processes targeted by this approach. If this is the case then there are grounds to believe that such approaches would be useful for those with MS, and potentially of greater benefit than approaches focused on pain control or reduction such as traditional CBT.

ACT and its mechanisms of change

ACT is one of the third generation behavioural therapies, differenti-ated from traditional behaviour therapy and cognitive-behavioural therapy by a greater focus on the context and functions of psycho-logical phenomena over their content or form (Hayes, 2004). Inter-ventions described as "contextual CBT" follow the same approach (McCracken et al., 2007). ACT does not aim to change the frequency or nature of any thoughts or feelings, as it is the response to thoughts and feelings that is seen as most important rather than how negative they are. One area where ACT differs most clearly from traditional CBT in this respect would be in that it does not use cognitive restruc-turing/disputation as a therapeutic technique. The emphasis in ther-apy is instead on changing how individuals respond to any given experience.

Therapeutic change in ACT is hypothesised to work to increase "psychological flexibility"—the ability to contact the present moment fully and without unnecessary attempts to control it, and at the same time to change or persist in behaviour in order to serve valued ends (Hayes et al., 2006). Key processes addressed with the aim of increas-ing psychological flexibility include experiential avoidance—the attempt to escape or avoid unwanted internal experiences (e.g., thought/worry) and cognitive fusion—the tendency for individuals to view thoughts as a true or accurate representation of reality (Hayes, 2004). These processes are not in themselves harmful, but become the focus of intervention if they are identified as getting in the way of living in a way the individual values. The term "acceptance" in this model is another aspect of psychological flexibility and is used to refer to the alternative to avoidance. It is the active taking in of what is afforded by the history and current context of the person. A related process that is distinguishable within the model is mindfulness, through which individuals attempt to increase awareness of psychological experiences in a way that is present focused, non-struggling, and non-evaluative (Vowles et al., 2009). Other key elements of the model include a focus on the client's values, or what is most important to them in terms of how they would choose to live their life, and committed action—the building of broad and flexible patterns of behaviour that work in service of these values. Although symptom reduction may occur, the aim is to explore and if necessary change how the person relates to their symptoms (e.g., thoughts and feelings associated with chronic pain).

A recent analysis of the empirical evidence concerning ACT found coherent support for the model, evidence for its efficacy across a broad range of psychological problems, and also concluded that the research suggests that it is working through its hypothesised process of change, that is, increasing psychological flexibility (Ruiz, 2010). This finding is supported by considerable research using the Acceptance and Action Questionnaire (AAQ) (Hayes et al., 2004). The AAQ was designed as a general measure of the ACT processes hypothesised to contribute to psychological flexibility and is probably the most frequently used measure in ACT process research. A meta analysis of thirty-two studies investigating the relationship between the AAQ and various quality of life outcomes found that on average psychological flexibility was correlated 0.42 with a diverse range of outcomes from job performance and satisfaction, mental health, and impact of pain on functioning. Versions of the AAQ in different contexts have found higher levels of psychological flexibility consistently associated with better quality of life and outcomes (Hayes et al., 2006).

Acceptance and psychological flexibility in MS and chronic pain

In a review of the empirical literature, Mohr and Cox (2001) outlined how behavioural/problem-focused coping strategies that attempt to alleviate problems that cannot be resolved may lead to frustration. The authors argue that, while the evidence to substantiate this is limited, this is a frequent occurrence in MS.

Cognitions are influential in adjustment, however, problematic cognitions in MS may be realistic and not amenable to challenges

(Dennison et al., 2009). For example, thoughts about symptoms worsening or becoming less independent may not necessarily be "distorted". Because of this, it may not be appropriate to target the content of cognitions as in traditional cognitive therapy.

Symptom reduction in MS is not always possible and commonly pain cannot be entirely eliminated or avoided. Treatment may mean a trade-off between pain and a loss of function (i.e., too much pain medication may minimise the pain but lead to increased fatigue, weakness, impact on cognition). While pain control is useful where possible (and can lead to improvements in functioning), attempts to fully control, reduce, or eliminate pain (and related thoughts/ emotions) may be unsuccessful in MS and continued attempts may be counterproductive. In such circumstances, acceptance (or "willingness" to experience psychological phenomena; the opposite of experiential avoidance) rather than control strategies may be more beneficial (Thompson & McCracken, 2011).

In the wider pain research literature evidence has been found for psychological flexibility as a mediator in improving functioning (Wicksell et al., 2010), and for experiential avoidance as a mediator between coping and psychopathology (Costa & Pinto-Gouvenia, 2011). In juvenile arthritis an investigation of the independent roles of pain intensity, psychological inflexibility, and acceptance of pain in predicting functional disability, anxiety, quality of life, and health related quality of life, greater psychological inflexibility has been found to uniquely predict higher anxiety, lower quality of life, and lower health related quality of life (Feinstein et al., 2011). Increases in acceptance of pain were found to be uniquely related to increase in quality of life. A correlational study also found that those chronic pain patients who demonstrated greater acceptance as measured with the Chronic Pain Acceptance Questionnaire (CPAQ) were the patients who used less health care resources and were the least distressed and disabled by their pain (McCracken et al., 2004).

While the theoretical argument for the potential benefits of treatments that target acceptance and psychological flexibility in chronic pain and MS is strong, more research is needed that looks specifically at the effects of targeting these psychological processes with this patient group.

Evidence for use of ACT with chronic pain

The literature specifically addressing ACT for chronic pain in MS remains limited and so it is useful to explore evidence arising from the use of ACT for pain in other chronic conditions. The American Psychological Association has recently acknowledged that the research support for the use of ACT with chronic pain is now "strong", and unlike other supported approaches, this is across "chronic and persistent pain in general" and not specific to a given pain syndrome (American Psychological Association, Division 12, 2011). Looking at individual studies, ACT has been demonstrated to be effective in reducing the impact of chronic pain on functioning in three large open trials in a specialist pain clinic setting (McCracken et al., 2005; McCracken et al., 2007; Vowles & McCracken, 2008). The first study used a contextual CBT three week interdisciplinary group treatment programme with 171 patients (Vowles & McCracken, 2008). Reliable change analysis suggested that three quarters of those treated demonstrated reliable improvement in depression, pain related anxiety, or overall disability at three month follow up, with the majority of these demonstrating reliable improvement in more than one domain. Changes in outcomes were related to changes of the hypothesised process measure: acceptance as measured by the CPAQ (McCracken et al., 2004) and values based action, as measured with the Chronic Pain Values Inventory (CPVI) (McCracken & Yang, 2006). There was no randomisation to a control condition, however, moderate to large improvements were seen at post treatment and follow up despite the long-standing, complex nature of participants' conditions. A follow up study found that significant improvements were still present at three years following treatment completion with 64.8% showing reliable improvements in at least one key domain. Improvements in acceptance and values based action were also found to be associated with improvements in emotional and physical functioning (Vowles et al., 2011).

The second study used a contextual approach with 108 patients with complex chronic pain (McCrackenet al., 2005). Comparison of within subjects measures of pain and functioning showed no significant changes during variable lengths of pre-treatment phase. Improved pain and functioning was, however, seen during the treatment phase and largely maintained at three month follow up. Effect sizes were large, with reductions in analgesic use and general practitioner (GP) visits also seen at follow up. Significant improvement in hypothesised process measures on the CPAQ were also seen during, but not before, the intervention. Changes in process measures also correlated highly with functional changes while changes in pain, although present, were comparatively small. This suggests that improvements were not solely due to a reduction in pain.

The third study employed a contextual approach with fifty-three highly disabled patients with chronic pain (McCracken et al., 2007). A clinical comparison group consisted of 234 adult patients with chronic pain, but without the level of disability, also completing a three week pain management programme. Statistically significant and clinically meaningful change was demonstrated for the highly disabled group including improvements in pain-related distress, physical and psychosocial disability, depression, pain related anxiety, daily rest due to pain, and acceptance of pain. Effect sizes were similar in magnitude to the clinical comparison group.

It should be noted that these three studies were all produced by the same team using a very intensive form of treatment delivery, typically six hours per day, five days per week, for three to four weeks. The psychological therapy is incorporated into an interdisciplinary approach including physiotherapists, occupational therapists, nurses, physicians, and clinical psychologists and the generalisability of these results to patients with MS seeking services in the community cannot be assumed.

There is however evidence from other sources to support the possibility of using less intensive ACT interventions outside a specialist setting. For example, a randomised controlled trial (RCT) comparing ACT and traditional CBT for chronic pain with primary care patients (N=114). Eight weekly group sessions were delivered, with improvements in pain interference, depression, and pain related anxiety found at six months post treatment (Wetherell et al., 2011). While no differences in outcomes were found between treatment groups, treatment completers were more satisfied with ACT. Data from two pilot studies also suggests that both eight and four session unidisciplinary ACT treatment groups for chronic pain are feasible, showing medium or large effects for acceptance, pain, and depression from pre- to post-treatment (Vowles et al., 2009). Effect sizes were comparable in both eight and four session interventions and were comparable, if not larger, than those seen for a five session CBT intervention that was delivered to a separate group.

Four small-scale randomised trials using ACT interventions with different groups have also shown promising results. The first showed improvements in functioning (demonstrated by reduced work absence and health care services use) for public sector workers showing chronic stress/pain and at risk of high sick leave utilisation (Dahl et al., 2004). The treatment group received four one-hour sessions of ACT in addition to treatment as usual. The second study was for chronic pain associated with whiplash in a specialist pain centre, which also showed significant improvements in disability, life satisfaction, and distress from ten sessions (Wicksell et al., 2008). In the third, (N = 70) outpatients attending an outpatient specialty pain clinic (average age of forty-six, 36% male, 64% female, reporting pain for over a year with 62% on sick leave) were randomised to seven week self-help manuals using either ACT or applied relaxation (AR). Both conditions had an initial face to face session, weekly therapist telephone support, and a face-to-face session at the end of treatment. Follow up at six and twelve months suggested better outcomes for ACT in level of function, pain intensity, acceptance, and marginal improvement in life satisfaction over the AR condition. Depression and anxiety improved significantly in both conditions but with no difference found between groups (Thorsell et al., 2011).

In the fourth study, ACT was delivered in a hospital setting to outpatients with chronic tension and chronic migraine type headache (Mo'tamedi et al., 2012). A reduction in disability and affective distress was demonstrated without significant changes in the reported sensory aspect of pain. This outcome is in line with ACT's focus on living well with pain rather than reducing pain directly.

There is a growing body of high quality evidence that supports the use of ACT for chronic pain across different causes and types, and that suggests it can positively affect several domains that are relevant to well-being. There is also evidence to support the feasibility of delivering ACT interventions for chronic pain in outpatient or community settings using relatively brief formats.

ACT and the wider challenges of MS

It is not just in relation to pain that ACT approaches may have something to offer people with MS. Increasing psychological flexibility may be of benefit in addressing the other challenges that people with the condition face. Chronic pain patients who report greater acceptance of negative experiences (not just pain related experiences) also report better social, emotional, and physical functioning (McCracken & Zhao-O'Brien, 2010). In theory, an approach focused on pain could have benefits for other areas. The nature of what people are accepting in relation to their own pain will be idiosyncratic (e.g., to include pain, guilt, thoughts of loss, images of failure) and so ideas of generalisability or specificity of treatment are not clear cut. The spectrum of what is to be struggled with or accepted is defined by the history and current context of the patient (McCracken & Zhao-O'Brien, 2010) not by the label of "pain".

A systematic review demonstrated that a range of psychological factors are associated with adjustment outcomes in MS (Dennison et al., 2009) with the strongest evidence found for a relationship between avoidant emotion-focused coping strategies and perceived stress relating to worse adjustment. Greater acceptance has also been found to be directly related to better adjustment to MS (Pakenham & Fleming, 2011). A systematic review and meta-analysis of acceptance based treatments for chronic pain indicated that acceptance based approaches such as mindfulness based stress reduction (MBSR) and ACT show small but equivalent effects to CBT on pain intensity and depression (Veehof et al., 2011). While the review concluded that mindfulness based approaches could be good alternatives to CBT, there are several reasons why ACT interventions could offer further benefits that were not identified by this review. Pain intensity was one of the primary outcomes in this review and while this is often affected by contextual treatment approaches it is neither the target of treatment nor the primary outcome. The majority of studies in this review were also MBSR, which does not contain the same focus on behaviour change as ACT. While there has been something of a paradigm shift from pain treatment focused on symptom reduction to approaches that include living well with pain (e.g., MBSR), approaches that more specifically address how skills such as mindfulness are integrated into daily life could have additional value. Individual ACT studies have reported large effects for depression (McCracken et al., 2007; Vowles & McCracken, 2008; Wicksell et al., 2008) and a medium sized effect for pain intensity (Vowles & McCracken, 2008). Using ACT interventions, large effects have also been found for pain related anxiety, sit to stand performance, and walking distance (McCracken et al., 2007; Vowles & McCracken, 2008).

There is a broad range of overlapping and interlinked symptoms and challenges in MS and it has been argued that ACT and mindfulness (which forms part of ACT) approaches have a broad scope and therefore potential for wider impact than just in one symptom domain such as fatigue or mood (Grossman et al., 2010; Mills & Allen, 2000; Sheppard et al., 2010). The potential usefulness of these interventions across a broad spectrum of difficulties is supported by studies where ACT has been applied successfully in the treatment of other chronic conditions such as reducing seizure frequency and increasing quality of life in drug refractory epilepsy (Lundgren et al., 2006; Lundgren et al., 2008), improving diabetes related self-care (Gregg et al., 2007), and reducing the interference and distress from tinnitus (Westin et al., 2011). This review found only two small studies looking at ACT for people with MS. The first, involving a one session workshop employing an ACT treatment protocol for psychosocial problems associated with MS, showed positive results with a small community sample (N = 15) at three month follow up (Sheppard et al., 2010). Significant improvements in depression, extent of thought suppression, impact of pain on behaviour, and quality of life were seen despite no change in the extent of physical symptoms. Large effects were seen for depression and impact of pain. There are however several methodological limitations that require the results to be interpreted with caution. Internal validity may have been compromised by the lack of a control group or condition so that it is harder to attribute observed effects to the treatment and it is not possible to differentiate any treatment effects from non-specific effects. Although not a specific target of the treatment, it is interesting that impact of pain was one area where change was seen. The authors acknowledge practical difficulties with engaging people with MS in long treatment sessions and suggest that future studies may look at multiple, shorter sessions.

The second study involved an RCT with twenty-one patients assigned to either ACT or relaxation training treatment conditions consisting of five sessions over fifteen weeks (Nordin & Rorsman, 2012). Patients showed no advantage of ACT on any of the outcome measures used in the study, however, the authors emphasised that the ACT treatment group showed a sustained improvement on a measure of acceptance (the AAQ-II) at three months follow up and both groups reported reductions in depressive symptoms. Interpretation of findings in this study were limited by the small sample size, the limited follow-up period, the absence of a waiting list control, and the lack of an independent evaluation of treatment integrity. Furthermore the study focused on measures of symptom reduction rather than behaviour change, which might have better represented the scope of an ACT approach.

The complexity and range of difficulties experienced by pwMS points to the need for a treatment approach that targets a range of psychological domains. While ACT's contextual approach provides a framework for addressing the management of chronic pain, the approach also addresses a broader range of psychological issues that may offer further useful benefits. The small body of research reviewed by this chapter offers limited but encouraging evidence in support of the use of ACT for pwMS, as well as evidence that ACT is operating in MS through its purported mechanisms of change.

Case study

The following is a common type of referral received by the community neurological service and an example of how ACT could be useful:

> Please see this young person who has a diagnosis of MS. They are low in mood and would benefit from psychological support

The assessment highlights the following: the patient, Tina was diagnosed with relapsing remitting MS five years ago. She is working as a manager in retail but feels that this is unsustainable due to experiencing a combination of fatigue, decline in mobility, and, most significantly, chronic neuropathic and musculoskeletal pain. She reports that pain is always present and not only affects her ability to concentrate at work, but also impacts on family relationships and stops her from being able to engage in many social activities. When she has recently engaged in previously enjoyed activities she tended to focus on her pain and has developed the belief that these activities make the pain worse, so it is better to withdraw from them. She feels low in mood, is no longer engaging in valued activities, and is spending more time alone. The more time Tina spends alone, the more she focuses on her pain, and the lower her mood becomes. She was previously prescribed baclofen but chose to stop using it as the side-effect of tiredness exacerbated her fatigue and led to concentration difficulties.

Tina was initially sceptical about psychological approaches but agreed to meet with the psychologist to see if there was anything else that could help her live with her MS symptoms. She was interested in ACT as an approach and agreed to attend eight sessions with the psychologist in the team.

In her sessions, the first stage involved exploring Tina's relationship with pain, what she had tried in dealing with it (medication, avoiding activities, "just putting up with it"), and what the short and long term outcomes of these different strategies had been. Tina described her relationship with pain as a "constant battle". She also found that constantly fighting something that did not go away was tiring and demoralising. As part of therapy Tina started developing her awareness of how she had been responding to pain and whether there were other possible ways of responding. Almost all of her responses to pain had been about moving away from pain through medication and restricting activity.

Tina found that despite five years of her best efforts to get rid of it the pain was still there. While some of her strategies for controlling pain were helpful at times, when she looked closely at her experiences she also found that sometimes her strategies did not provide any further relief from pain, but they did take something else away from her life (e.g., avoiding seeing friends did not get rid of the pain and she missed out on enjoying their company). Although she had not wanted to think about it before, it did not seem like doing more of the same thing was going to get rid of the pain. Tina decided that it was not possible for her to ultimately control or eliminate pain completely, but that in attempting to control something that could not be controlled, her suffering increased and her life became more restricted.

She started practising "acceptance"—being genuinely willing to engage fully in an activity whether pain is present or not. This was introduced through experiential exercises in sessions then practised at home. The pain was still there, but she reported increased satisfaction with her life from engaging in valued activity and an improvement in her mood.

Tina often had critical and self-defeating thoughts associated with pain, which, more often than not, she acted on. By practising "de-fusing" from thoughts (being able to separate herself from her thoughts and view them just as thoughts rather than challenging them or disputing them) Tina found that while she still had the same thoughts about pain they were not dominating her behaviour. For example, a recurring thought was "the pain will be unbearable, don't do it", which she initially treated as an unbreakable rule. She practised treating this as a thought about pain rather than a fact about pain (again through experiential exercises in sessions to begin with) and found that she was developing confidence at being able to engage in valued activities whether difficult thoughts showed up or not.

In sessions Tina had also started practising exercises aimed at bringing her attention to the present moment, as opposed to what she noticed she had often been doing, which was responding to past hurt or a feared future. Importantly, she had also started transferring this practise to everyday situations outside sessions.

Another facet of the treatment approach was developing an awareness of the "self as context" or "observer self" through engaging in exercises that allowed her to experience this perspective. She said that she often found herself getting "tangled up" in her experiences, and was starting to be able to take a step back from them, recognising that whether they were positive or negative, there was an ongoing stream of experience, and that these experiences did not define her.

A vital aspect of the intervention was spending time focusing on Tina's values. This essentially involved attempting to answer the question "what do you want your life to be about?". This is about the quality she wanted to bring to interactions rather than specific goals. For her a key value was "engaging openly with friends". She found that it was possible to bring this quality to her interactions with friends even when she could not always do the things she used to and even when pain was present. Setting goals, identifying psychological barriers that showed up in pursuing her goals, and using the skills developed in therapy to find ways to move forwards with these barriers were ongoing aspects of therapy. The overall goal was building broader patterns of more flexible behaviour. At the end of therapy Tina was still reporting significant levels of pain, however, the impact of the pain on her functioning had reduced, in particular she was seeing friends more, whether or not she was still experiencing pain. She said that some aspects of the treatment had been difficult, for example, taking a step back and noticing her feelings about the pain made her realise that she had been ignoring how scared she was about her symptoms worsening. Recognising these feelings had not made her feel better, but had allowed her to talk to friends about it, which she had found supportive, and she also recognised that this had been an important part of what she described as starting to be "kinder to herself". She said that she felt relieved when the focus of treatment was on what she wanted and valued in her life as she said that it had seemed like her life was becoming a list of symptoms and that she felt her identity was getting lost in this. Follow up after three months showed that Tina had incorporated aspects of the treatment sessions into everyday practice, an improvement in her mood had been maintained, and the impact of pain on her functioning had reduced significantly.

Conclusion and implications

The evidence to support the use of ACT interventions for chronic pain in MS is currently limited. However, there is evidence to support the usefulness of ACT with depression and adjustment in MS as well as wider evidence that contextual approaches are of particular benefit in supporting people to live well with pain in other chronic conditions. ACT may offer an advantage over traditional CBT as its therapeutic processes target multiple relevant domains. The evidence for the use of ACT processes in chronic pain and with other chronic conditions is developing rapidly and supports the need for further experimental research on the effectiveness of this intervention for chronic pain in MS. In order to determine whether ACT may be of benefit to people with MS and chronic pain, studies also need to address the feasibility of this approach in clinical settings. It has been possible to deliver effective ACT interventions for other problems in a unidisciplinary, brief format. It would be worthwhile exploring this design in MS and chronic pain as this is likely to be the most feasible format for community services to attempt to deliver.

Psychological flexibility requires attention, executive functions, and working memory (Kashdan & Rottenberg, 2010); areas that may be impaired in MS. This could pose a challenge for an intervention whose success purports to depend on developing this capacity. The available evidence on this is unclear, however, with some authors discussing the possible advantages of the use of metaphors in ACT when working with people who have impaired working memory or slow processing (Kangas & McDonald, 2011) or with other organic deficits (Ylvisaker et al., 2008). It is beyond the scope of the current review to explore this further, however, this is another area worth exploring in order to help clinicians decide which interventions are likely to be most beneficial for which patients.

While the current lack of high quality research using ACT with this patient group means there are still fundamental questions about the usefulness of this approach, the available evidence supports the value of further research including small scale clinical studies of ACT for chronic pain in MS. Initial questions to be answered should include whether the approach can be delivered in a realistic community neurological service/outpatient setting, and whether it is acceptable to patients. Collecting data on treatment process, as well as broad outcomes that reflect quality of life not just frequency/intensity of pain or other symptoms, will help identify what the impact of ACT treatment processes is for pwMS, and any interaction with patient characteristics. Collection of data on cognitive functioning would also allow for the exploration of the impact of cognitive problems associated with MS on the ability to benefit from this approach. Once questions around feasibility, acceptability, and effectiveness have been answered, it should be established whether or not ACT has anything additional to offer over traditional CBT.

References

American Psychological Association, Division 12 (2011). *Chronic or Persistent Pain in General (Including Numerous Conditions)*. Available at: www.div12.org/PsychologicalTreatments/disorders/pain_general.php. Accessed 12 January 2012.

Costa, J., & Pinto-Gouveia, J. (2011). The mediation effect of experiential avoidance between coping and psychopathology in chronic pain. *Clinical Psychology and Psychotherapy, 18*: 34–47.

Dahl, J., Wilson, K. G., & Nilsson, A. (2004). Acceptance and commitment therapy and the treatment of persons at risk of long-term disability resulting from stress and pain symptoms: a preliminary randomised trial. *Behavior Therapy, 35*: 785–801.

Dennison, L., Moss-Morris, R., & Chalder, T. (2009). A review of psychological correlates of adjustment in patients with multiple sclerosis. *Clinical Psychology Review, 29*: 141–153.

Feinstein, A. B., Forman, E., Masuda, M., Cohen, L. L., Herbert, J. D., Moorthy, L. N., & Goldsmith, D. P. (2011). Pain intensity, psychological inflexibility, and acceptance of pain as predictors of functioning in adolescents with juvenile idiopathic arthritis: a preliminary investigation. *Journal of Clinical Psychology in Medical Settings, 18*(3): 291–298.

Foley, J. F., & Brandes, D. W. (2009). Redefining functionality and treatment efficacy in multiple sclerosis. *Neurology, 72*: s1–s11.

Forbes, A., While, A., Mathes, L., & Griffiths, P. (2006). Health problems and health related quality of life in people with multiple sclerosis. *Clinical Rehabilitation, 20*: 67–78.

Gregg, J. A., Callaghan, G. M., Hayes, S. C., & Glenn-Lawson, J. L. (2007). Improving diabetes self-management through acceptance, mindfulness and values: a randomised controlled trial. *Journal of Consulting and Clinical Psychology, 75*(2): 336–343.

Grossman, P., Kappos, L., Geniscke, H., D'Souza, M., Mohr, D. C., Penner, I. K., & Steiner, C. (2010). MS quality of life, depression and fatigue improve after mindfulness training: a randomised trial. *Neurology, 75*: 1141–1149.

Hawthorne, G., Richardson, J., & Osbourne, R. (1999). The assessment of quality of life (AQoL) instrument: a psychometric measure of health related quality of life. *Quality of Life Research, 8*: 209–224.

Hayes, S. C. (2004). Acceptance and commitment therapy, relational frame theory, and the third wave of behavioural and cognitive therapies. *Behaviour Therapy, 35*: 639–665.

Hayes, S. C., Luoma, J., Bond, F. W., Masuda, A., & Lillis, J. (2006). Acceptance and commitment therapy: model, process and outcomes. *Behaviour Research and Therapy, 44*: 1–25.

Hayes, S. C., Strosahl, K. D., Wilson, K. G., Bissett, R. T., Pistorello, J., & Toarmino, D. (2004). Measuring experiential avoidance: a preliminary test of a working model. *The Psychological Record, 54*: 533–578.

Heckman-Stone, C., & Stone, C. (2001). Pain management techniques used by patients with multiple sclerosis. *The Journal of Pain, 2*(4): 205–208.

Kangas, M., & McDonald, S. (2011). Is it time to act? The potential of acceptance and commitment therapy for psychological problems following acquired brain injury. *Neuropsychological Rehabilitation, 21*(2): 250–276.

Kashdan, T. B., & Rottenberg, J. (2010). Psychological flexibility as a fundamental aspect of health. *Clinical Psychology Review, 30*: 865–878.

Khan, F., & Pallant, J. (2007). Chronic pain in multiple sclerosis: prevalence, characteristics, and impact on quality of life in an Australian community cohort. *The Journal of Pain, 8*: 614–623.

Lundgren, T., Dahl, J. A., Melin, L., & Kies, B. (2006). Evaluation of acceptance and commitment therapy for drug refractory epilepsy: a randomised controlled trial in South Africa—a pilot study. *Epilepsia*, 47(12): 2173–2179.

Lundgren, T., Dahl, J., Yardi, N., & Melin, L. (2008). Acceptance and commitment therapy and yoga for drug-refractory epilepsy: a randomised controlled trial. *Epilepsy and Behaviour*, 13: 102–108.

McCracken, L. M., & Gutierrez-Martinez, O. (2011). Processes of change in psychological flexibility in an interdisciplinary group-based treatment for chronic pain based on acceptance and commitment therapy. *Behaviour Research and Therapy*, 49(4): 267–274.

McCracken, L. M., & Yang, S. (2006). The role of values in a contextual cognitive–behavioral approach to chronic pain. *Pain*, 123: 137–145.

McCracken, L. M., & Zhao-O'Brien, J. (2010). General psychological acceptance and chronic pain: there is more to accept than the pain itself. *European Journal of Pain*, 14: 170–175.

McCracken, L. M., MacKichan, F., & Eccleston, C. (2007). Contextual cognitive–behavioral therapy for severely disabled chronic pain sufferers: effectiveness and clinically significant change. *European Journal of Pain*, 11: 314–322.

McCracken, L. M., Vowles, K. E., & Ecclestone, C. (2004). Acceptance of chronic pain: component analysis and a revised assessment method. *Pain*, 107:159–166.

McCracken, L. M., Vowles, K. E., & Eccleston, C. (2005). Acceptance-based treatment for persons with complex, longstanding chronic pain: a preliminary analysis of treatment outcome in comparison to waiting phase. *Behavior Research and Therapy*, 43: 1335–1346.

Miller, A. E. (2001). Clinical features. In: S. D. Cook (Ed.), *Handbook of Multiple Sclerosis* (3rd edn). New York: Marcel Dekker.

Mills, N., & Allen, J. (2000). Mindfulness of movement as a coping strategy in multiple sclerosis: a pilot study. *General Hospital Psychiatry*, 22(6): 425–431.

Mohr, D. C., & Cox, D. (2001). Multiple sclerosis: empirical literature for the clinical health psychologist. *Journal of Clinical Psychology*, 57(4): 479–499.

Mo'tamedi, H., Rezaiemaram, P., & Tavallaie, A. (2012). The effectiveness of a group-based acceptance and commitment additive therapy on rehabilitation of female outpatients with chronic headache: preliminary findings reducing 3 dimensions of headache impact. Headache: *The Journal of Head and Face Pain*, 52(7): 1106–1119. doi:10.1111/j.1526-4610.2012.02192.x.

Moulin, D. E., Foley, K. M., & Ebers, G. C. (1988). Pain syndromes in multiple sclerosis. *Neurology, 38*(12): 1830.

Multiple Sclerosis Trust. (2008). *MS Explained*. Available at: www.mstrust. org. uk/shop/product.jsp?prodid=84. Accessed 7 May 2011.

Nordin, L., & Rorsman, I. (2012). Cognitive behavioural therapy in multiple sclerosis: a randomized controlled pilot study of Acceptance and Commitment Therapy. *Journal of Rehabilitative Medicine, 44*: 87–90.

O'Connor, A., Schwid, S. R., Hermann, D. N., Markman, J. D., & Dworkin, R. H. (2008). Pain associated with multiple sclerosis: systematic review and proposed classification. *Pain, 137*: 96–111.

Pakenham, K. I., & Fleming, M. (2011). Relations between acceptance of multiple sclerosis and positive and negative adjustments. *Psychology and Health, 26*(10): 1292–1309.

Richards, R. G., Sampson, F. C., Beard, S. M., & Tappenden, P. (2002). A review of the natural history and epidemiology of multiple sclerosis: implications for resource allocation and health economic models. *Health Technology Assessment, 6*(10): 1–73.

Rossi, S., Mataluni, G., Codeca, C., Fiore, S., Buttari, F., Musella, A., Castelli, M., Bernardi, G., & Centonze, D. (2009). Effects of levetiracetam on chronic pain in multiple sclerosis: results of a pilot, randomised, placebo-controlled study. *European Journal of Neurology, 16*: 360–366.

Ruiz, F. J. (2010). A review of acceptance and commitment therapy (ACT) empirical evidence: correlational, experimental, psychopathology, component and outcome studies. *International Journal of Psychology and Psychological Therapy, 10*(1): 125–162.

Sadovnik, A. D., Ebers, G. C., Wilson, R. W., & Paty, D. W. (1992). Life expectancy in patients attending multiple sclerosis clinics. *Neurology, 42*: 991–994.

Sheppard, S., Forsyth, J., Hickling, E. J., & Bianchi, J. M. (2010). A novel application of acceptance and commitment therapy for psychosocial problems associated with multiple sclerosis: results from a half day workshop. *International Journal of MS Care, 12*: 200–206.

Solaro, C., & Uccelli, M. M. (2011). Management of pain in multiple sclerosis, a pharmacological approach. *Nature Reviews Neurology, 7*(9): 519–527.

Stewart, A. L., & Ware, J. E. (Eds.). (1992). *Measuring Functioning and Well-being: The Medical Outcomes Study Approach*. Santa Monica, CA: RAND Corporation.

Thomas, P. W., Thomas, S., Hillier, C., Galvin, K., & Baker, R. (2006). Psychological interventions for multiple sclerosis. *Cochrane Database of*

Systematic Reviews, (1). Available at: http://www.update-software.com/pdf/CD004431.pdf

Thompson, M., & McCracken, L. M. (2011). Acceptance and related processes in adjustment to chronic pain. *Current Pain and Headache Reports*, *15*: 144–151.

Thorsell, J., Finnes, A., Dahl, J., Lundgren, T., Gybrant, M., Gordh, T., & Buhrman, M. (2011). A comparative study of 2 manual based self-help interventions, acceptance and commitment therapy and applied relaxation, for persons with chronic pain. *Clinical Journal of Pain, 27*(8): 716–723.

Veehof, M. M., Oskam, M. J., Schreurs, K. M., & Bohlmeijer, E. T. (2011). Acceptance-based interventions for the treatment of chronic pain: a systematic reveiw and meta-analysis. *Pain, 152*(3): 533–542.

Vowles, K. E., & McCracken, L. M. (2008). Acceptance and values-based action in chronic pain: a study of treatment effectiveness and process. *Journal of Consulting and Clinical Psychology, 76*: 497–507.

Vowles, K. E., Loebach Wetherell, J., & Sorrell, J. T. (2009). Targeting acceptance, mindfulness, and values based action in chronic pain: findings of two preliminary trials of an outpatient group-based intervention. *Cognitive and Behavioural Practice, 16*: 49–58.

Vowles, K. E., McCracken, L. M., & O'Brien, J. Z. (2011). Acceptance and values-based action in chronic pain: a three-year follow-up analysis of treatment effectiveness and process. *Behaviour Research and Therapy, 49*(11): 748–755.

Westin, V. Z., Schulin, M., Hesser, H., Karlsson, M., Noe, R. Z., Olofsson, U., Stalby, M., Wisung, G., & Andersson, G. (2011). Acceptance and commitment therapy versus tinnitus retraining therapy in the treatment of tinnitus: a randomised controlled trial. *Behaviour Research and Therapy, 49*(11): 737–747.

Wetherell, J. L., Afari, N., Rutledge, T., Sorrell, J. T., Stoddard, J. A., Petkus, A. J., Solomon, B. C., Lehman, D. H., Liu, L., Lang, A. J., & Atkinson, J. H. (2011). A randomised controlled trial of acceptance and commitment therapy and cognitive–behavioral therapy for chronic pain. *Pain, 152*(9): 2098–2107.

Whetten-Goldstein, K., Sloan, F. A., Goldstein, L. B., & Kulas, E. D. (1998). A comprehensive assessment of the cost of multiple sclerosis in the United States. *Multiple Sclerosis, 4*: 419–425.

Wicksell, R. K., Ahlquist, J., Bring, A., Melin, L., & Olson, G. L. (2008). Can exposure and acceptance strategies improve functioning and life satisfaction in people with chronic pain and whiplash associated disorders

(WAD)? A randomised controlled trial. *Cognitive Behavior Therapy, 37*: 169–182.

Wicksell, R. K., Olsson, G. L., & Hayes, S. C. (2010). Psychological flexibility as a mediator of improvement in acceptance and committment therapy for patients with chronic pain following whiplash. *European Journal of Pain, 14*: 1059.e1–1059.e11.

Ylvisaker, M., Mcpherson, K., Kayes, N., & Pellett, E. (2008). Metaphoric identity mapping: facilitating goal setting and engagement in rehabilitation after traumatic brain injury. *Neuropsychological Rehabilitation, 18*(5–6): 713–741.

CHAPTER TWO

Acceptance and commitment therapy for the management of chronic neuropathic pain in multiple sclerosis: a case study*

Neil Carrigan and Leon Dysch

Introduction

Multiple sclerosis (MS) is a neurological disease that is caused by demyelination of nerve fibres in the brain and spinal cord. The disease is progressive with damage disrupting communication between parts of the central nervous system, leading to both physical and cognitive symptoms that may include visual impairment, fatigue, ataxia, pain, numbness, and cognitive decline. Around 0.1% of people in the UK will be affected by MS (Richards et al., 2002). Chronic pain in MS is relatively common with an estimated 48–65% experiencing some form of long-standing pain (Khan & Pallant, 2007; Moulin et al., 1988; O'Connor et al., 2008). In MS, chronic pain has been shown to have a negative impact on a range of measures including health related quality of life, psychological well-being, and independent living (Forbes et al., 2006; Hawthorne et al., 1999; Sherbourne, 1992). Hence the need for effective treatments that help patients manage their pain.

One form of chronic pain in MS is central neuropathic pain, defined as "pain caused by a lesion or disease of the central somatosensory

---

* Originally published in 2015 in *Neuro-Disability & Psychotherapy*, 3(2): 69–92.

nervous system" (IASP, 2007, p. 1). It affects approximately 26% of people with MS (Foley et al., 2013). Neuropathic pain can be constant or intermittent with the onset of pain episodes happening spontaneously or in response to a stimulus (NICE, 2013). People suffering from neuropathic pain often use words such as "stabbing", "shooting", "electric shock", "burning", "tingling", "pins and needles", to describe the pain (IASP, 2007). The exact cause and mechanisms of neuropathic pain are often difficult to discern, making its management by health professionals problematic (Beniczky et al., 2005). The recommended pharmacological approach to treating neuropathic pain includes antidepressants, antiepileptic drugs, topical treatments, and opioids (NICE, 2013). While a pharmacological approach is often used to manage neuropathic pain in MS (Heckman-Stone & Stone, 2001), many of the drugs are not licensed for use in this way (which may limit their use) and their potential benefits are often outweighed by their adverse side-effects (NICE, 2013), for example, drowsiness. There are also perceived barriers to accessing pain treatment including lack of accessible pain or MS specialists, fear of taking medication/dependence, lack of finances, and the belief that nothing works for pain (Khan & Pallant, 2007). So while approximately 30% of all medications prescribed to manage symptoms in MS patients relate to pain (Solaro & Uccelli, 2011) the medical approach to its management is still seen as far from satisfactory (Rossi et al., 2009).

A further difficulty in adopting a purely medical approach to pain management is that research is now demonstrating that long-term pain comprises an affective component as much as a sensory one (IASP, 2007). This affective component of pain comprises an emotional evaluation of the pain in terms of how unpleasant it feels to the patient and the amount of suffering it causes. It is seen as distinct from the sensory quality of the pain (pain intensity, etc.) and contributes significantly to the overall pain experience (Reading et al., 1982; Turk et al., 1985). Hence any approach to the management of chronic pain in MS needs to acknowledge the psychological factors that contribute to it. While the UK National Institute for Clinical Health and Excellence (NICE) recommends psychological therapy in the management of neuropathic pain, they do not as yet specify a particular form of therapy (NICE, 2013).

By far the most studied psychological treatment used in chronic pain is cognitive behaviour therapy (CBT) (Williams et al., 2012). The most recent systematic review of randomised controlled trials (RCTs) of CBT for chronic pain found small to moderate effect sizes on measures of pain, disability, mood, and catastrophising when compared to treatment as usual or wait list controls but only small effect sizes on these measures when compared to active control groups (Williams et al., 2012). However, these improvements disappeared at follow-up leading the authors to conclude that the benefits of CBT are found mainly in comparisons with waiting list and treatment as usual, rather than active controls. In reviewing these in a recent meta-analyses, Morley and colleagues (2013) noted that half of the studies found no effect for CBT while the other half found only weak effects sizes whose significance in terms of clinical outcome were unclear. They go on to suggest that while CBT for chronic pain is likely to be effective, further studies are needed that pinpoint which elements of CBT are likely to be effective for which patients and why (Williams et al., 2012).

Recent developments in CBT, known as "third-wave", have shown early promise in helping patients overcome their problems with chronic pain. One such third-wave approach is acceptance and commitment therapy (ACT) (Hayes et al., 2012). ACT is derived from relational frame theory (RFT), which is seen as a theory of human language and cognition based on post-Skinnerian behavioural principles (Hayes et al., 2001). According to the theory, psychological distress is caused by: "the way that language and cognition interact with direct contingencies to produce an inability to persist or change behaviour in the service of long term valued ends" (Hayes et al., 2006, p. 6). The main mechanism by which distress arises is from a lack of psychological flexibility that results from six interdependent processes (see Figure 1). In ACT, words and images can function as if they are the actual events they describe and this "cognitive fusion" (in Figure 1) of thoughts with reality can prevent the use of potentially more adaptive and useful sources of information to regulate behaviour, such as direct experience of the world as perceived through the senses ("contact with present moment"). Hence problems arise when people do not see themselves as separate from the process of thinking ("self-as-context") or trying to manage their distress by attempting to escape or avoid unpleasant private events such as thoughts and feelings (lack of "committed action"). In doing so they may behave in ways that contradict their "values" and hamper their achievement of valued goals. ACT/RFT emphasises the difference between the form and function of a behaviour as well as the

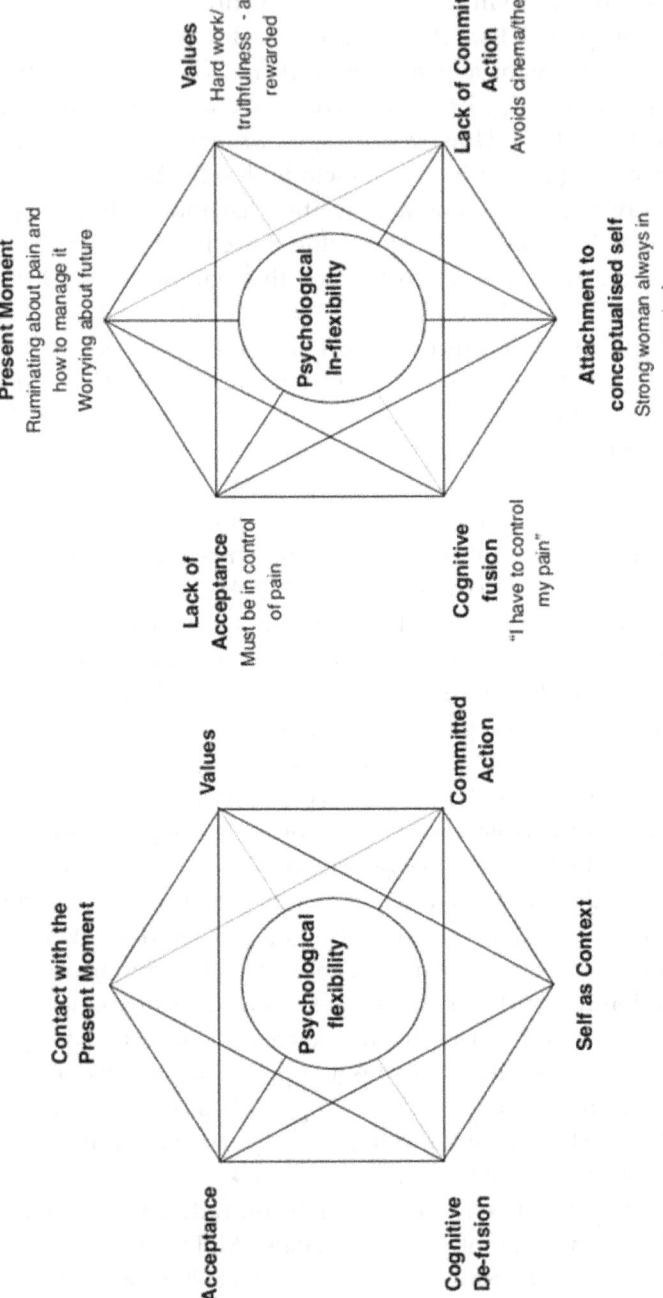

Figure 1: W's hexaflex model of psychological flexibility (left) with a hexaflex of psychological inflexibility (right) (taken from Hayes et al. (2012)).

context in which it occurs. So, for example, panic is not seen as just a more intense form of anxiety, but that the behavioural and psychological function of panic is to avoid and resist feared situations. Similarly in chronic pain, it is not just pain intensity that maintains distress but "experiential avoidance" (Hayes & Duckworth, 2006).

Cognitive therapeutic approaches generally see distress caused by long term pain as maladaptive and seek to alleviate this distress through helping the patient to challenge the thoughts, beliefs, and meanings attributed to their pain that are maintaining their distress (a process known as cognitive restructuring). Some have argued that such an approach is unlikely to succeed for patients with chronic long term pain (and others with chronic illness/conditions) because their beliefs may be accurate reflections of their objective reality (Graham et al., 2015; Harrison et al., 2014). While cognitive therapists might argue that instead of tackling such "reality cognitions", a more useful target would be the personal meanings associated with the condition, doing so requires careful teasing apart of reality from the beliefs the patient attributes to that reality. This may prove difficult for patients with long term pain where it has become an established part of their daily lives, or when therapists have only basic training in cognitive behavioural approaches.

In contrast, ACT eschews the problems of tackling and challenging distressing thoughts about pain directly and instead asks the patient to give up trying to control and change experience. Instead it encourages psychological flexibility by helping patients become more accepting of current psychological states and engage in behaviour that moves them towards more meaningful and valued ends (Hayes et al., 2006). Thus the aim of ACT is to change the relationship a person has towards their symptoms and not the reduction of symptoms per-se (although this may occur). It uses techniques such as mindfulness meditation to promote a non-struggling, non-evaluative acceptance of the present moment and awareness of the current context afforded to the person. ACT also seeks to clarify the person's values and encourage them to commit to behave in ways that serve these values and move towards a more meaningful life (Hayes et al., 2012). For a more detailed account of the theory and philosophical underpinnings of ACT see Hayes and colleagues (2012).

ACT has been shown across a growing number of trials to improve outcomes in heterogeneous chronic pain samples (APA, 2014). One

RCT that compared ACT with treatment as usual found significant improvements in the ACT arm of the study on measures of pain disability, life satisfaction, fear of movements, depression, and psychological flexibility compared to treatment as usual—but no difference in level of pain intensity (Wicksell et al., 2008). This is consistent with other chronic pain treatment programmes using ACT that suggest improvements in outcomes are not necessarily dependent on changes in pain symptoms but on acceptance of pain and the willingness to act towards achieving one's goals despite its presence (McCracken & Gutiérrez-Martínez, 2011; McCracken et al., 2004; Wicksell et al., 2010). A later RCT compared ACT against CBT and while they found no difference between the two groups on any of the treatment outcome variables, ACT participants reported significantly greater satisfaction with treatment compared to the CBT participants (Wetherell et al., 2011). The finding suggests ACT can be offered as a viable alternative to standard CBT giving patients choice in the psychological approach they want in tackling their chronic pain. Given that psychological flexibility requires attendance to factors that can influence behaviour (such as thoughts, feelings, and the context in which the behaviour takes place), as well as being able to inhibit undesired behaviour, there is concern that patients suffering from neurological conditions may not have the cognitive faculties to engage in ACT. However, there is a small but growing literature demonstrating the benefit of using ACT with patients experiencing neurological impairment. For example, Sylvester (2011) describes an eight session ACT group for adults who suffered acquired brain injury during childhood. The group led to greater engagement in helpful behaviours and reduced avoidance. In their review of the literature, Kangas and MacDonald (2011) suggest ACT can be effective in this patient population and there is a hope of more definitive evidence from the results of a randomised controlled trial of ACT for traumatic brain injury currently being conducted by Hamish McLeod and colleagues (Whiting et al., 2012). People with MS face a number of challenges that impact on physical, psychological and emotional functioning, and overall quality of life. There are few studies at present where ACT has been used for people with MS. Sheppard and colleagues (2010) provided MS patients with a half-day workshop on ACT and found that while it did not improve physical symptoms (as would be expected) it did significantly improve depression rates, the impact of pain on behaviour, and quality of life at three month follow up. Gillanders and Gillanders (2014) report on a sixty-two-year-old woman with secondary progressive MS whose thoughts about her MS as well as childhood trauma were having a negative impact on her psychological well-being. Following ten sessions of individual ACT treatment over a six month period her anxiety and depression fell to non-clinical levels, as well as showing greater acceptance of her MS and engagement in valued activities. Tooze and colleagues (unpublished) piloted an ACT intervention for chronic pain in twelve MS patients using a group format given over four, two-hour sessions. They found a significant increase in the acceptance of pain at treatment completion but no differences in pain intensity, the impact of pain, or in psychological flexibility. The reasons given for a lack of hypothesised effect on the last two variables was the brief nature of the intervention and the lack of experience of the group facilitators in using the ACT approach. However, self-report measures suggested that patients perceived the treatment as credible, effective, and relevant to the problems they faced as MS patients suffering chronic pain.

A recent paper by Tooze and colleagues (2014) explored the potential benefits of using ACT for chronic pain in MS. However, as yet there are no (to the authors' knowledge) published individual case studies of the use of ACT in treating chronic pain in MS. The case presented here reports the successful treatment of a woman with MS suffering with neuropathic pain who approached her pain specialist specifically looking for a mindfulness based approach to "control" her pain. The treatment followed the same protocol developed by Tooze and colleagues (unpublished) with minor adaptations for an individual rather than a group. Given the findings from previous trials of ACT with chronic pain, it was expected that the patient would show improvement on measures of psychological well-being without necessarily showing a reduction in pain symptoms.

Method

Case history

At the time of referral W was a sixty-two-year-old woman who had been diagnosed with MS in 1993. She was a retired teacher but was still working part-time in a similar role and lived with her husband. She had grown up children who lived separately to W and her husband but two of her children were experiencing distress from physical illness in their respective families and would often call on W for advice or to talk about their problems. W's husband was support-ive of her and would bring her to appointments (although did not attend the sessions). Since 1993 her MS had affected her balance with mild spasticity that had led to several significant falls that caused frac-tures. For a number of years she had been suffering with neuropathic pain that she described as a sharp pain or heat in her lower legs and hands that became worse towards the end of the day. Her main method of managing her pain had been through medication. Her pain medication included paracetamol, amitriptyline, and gabapentin. W was also taking propranolol for hypertension. There were no changes to her medication prior to or during the six week ACT intervention. She felt that while her medication did provide some relief it was dependent on taking it before the pain became problematic but this did not work every time. On occasion W would rest in the early evening in an attempt to mitigate the pain and this led to her reduc-ing her activity levels. She had tried very hard to understand what might trigger her pain but an understanding of what exacerbated or mitigated it had so far eluded her. Her inability to control her pain had left her feeling very frustrated. She recognised that this frustration in turn added to the distress caused by her pain.

Cognitively she experienced word finding difficulties and she reported that her memory was not as sharp as it used to be. However, as she did not report any functional impairment from these difficul-ties, it was not thought appropriate to put her through a lengthy cognitive assessment. Her mood was generally positive but occasion-ally there were days where her mood was low, but not seen as prob-lematic. She felt anxious about falling and while she tried to prevent this from stopping her doing anything, it increased the distress she felt when engaging in physical activities. She did not feel that mood or anxiety were problematic and so were not formally assessed.

W reported that when in pain, she could not keep her feet still and needed to stand up and move around. This caused difficulties when engaging in social activities such as going to the cinema or theatre. The pain could also affect her concentration, her work, and could affect her driving. At her initial assessment she stated that her pain had become increasingly difficult to manage. This was causing her distress as she saw herself as an intelligent, strong, independent woman. Hence being able to drive, attend the theatre/cinema, and perform well at work contributed to what she valued in her life. Her frustrations with her inability to control her pain and thus engage in these activities had led to a subjective lowering of her quality of life. As a result she asked for help from her pain consultant as she desperately wanted to "control" her relationship with pain. It was at her initiation, and following discussion with her consultant, that she was referred to the community neuro team for a mindfulness-based approach to pain management because she no longer felt that medication, or her own attempts to control her pain, were successful.

W was familiar with meditative approaches and very willing to engage in treatment although more reluctant about committing to valued actions as she felt she was already doing this.

Formulation

In ACT, the main hypothesis is that a person's distress is due to a lack of psychological flexibility in managing their difficulties. Figure 1 presents the "hexaflex" model (Hayes et al., 2012) traditionally used to demonstrate how these different areas interact to cause psychological inflexibility (alongside a hexaflex model of psychological flexibility). W had become quite inflexible in her approach to her pain. She naturally wanted to control her pain and her thinking had become "cognitively fused" with the thought "I have to control my pain" as though this was a literal truth. She had become increasingly distressed that she was unable to do so. She saw herself as a strong independent woman who was very much in control of every other area of her life and did not understand why she could not control her pain.

W had been a successful teacher and had worked hard for her success. She had strong values around fairness and honesty: if you work hard and are truthful then you should be rewarded. In her work life this had been the case and while she tried to not let pain prevent her from doing activities, she was frustrated at how much it interfered with what she valued in life. In terms of Figure 1, due to these values she had come to view her pain as "unfair" as it was unpredictable and occurred even after she had done all in her power to alleviate it. The lack of perceived fairness and the uncontrollability of her pain had challenged her view of herself as someone in control (self as context) and left her feeling frustrated and distressed. She could not distance herself from this view of herself as a person always in control or accept that she actually had very little control over her pain (acceptance). While she did many things, she avoided activities that required sitting for long periods such as going to the cinema or theatre, as she thought she would be unable to sit in her seat for the duration because of her pain (committed action). W spent a lot of time thinking about her pain or worrying about how to manage her pain in the future, both in the short term (later the same day) and long term (whether her pain would get worse in the future), and thus spent little time in contact with the present moment.

Design

The case study follows an AB (assessment followed by treatment) design (Barlow & Hersen, 1984). The patient was assessed two weeks before treatment commenced and completed a series of baseline measures. Process measures were taken weekly throughout the six session treatment. The same measures used at baseline were taken post-treatment and at two month follow-up. It was hypothesised that following treatment W would show improvements on measures of quality of life (QLI), the impact of pain (PES), and fatigue (MFIS-5) but it was not expected that her level of pain intensity (BPI) would necessarily improve given the previous research reviewed above.

Measures

The following measures were taken as a baseline at assessment and repeated at the end of treatment:

> *Quality of life index (MS Version)* (QLI) (Ferrans & Powers, 2007). This measures quality of life in terms of how satisfied the participant is with different areas of their life, and also how important the participant rates each of these areas. Scores range from zero to thirty. The QLI has been used in studies of various physical health conditions (including MS) demonstrating good levels of reliability ($\sim = 0.79$) and validity (Stuifbergen, 1995).
>
> *Medical outcomes study pain effects scale* (PES) (Sherbourne, 1992) is a six item self-report measure of the impact of pain on functioning. Scores indicate the extent to which symptoms have interfered with mood, mobility, sleep, work, recreational activities, and

enjoyment of life over the past four weeks. Possible scores range from six to thirty; higher scores indicate greater impact of pain. The scale has good internal consistency (0.86) (Ritvo et al., 1997). *Modified fatigue impact scale* (MFIS-5) is a fatigue sub-scale of the MS quality of life inventory (Ritvo et al., 1997). It is a five item self-report measure assessing the impact of fatigue across five domains including alertness, concentration, and ability to complete physical tasks. Scores can range from zero to twenty; higher scores suggesting a greater negative impact of fatigue. The scale's Cronbach's alpha is 0.8.

At every treatment session the following measures were taken to track her progress:

Chronic pain acceptance questionnaire (CPAQ) (McCracken et al., 2004). This is a twenty item measure of overall acceptance of pain that includes two components: activities engagement and pain willingness. The first measures the degree to which pain and related experiences restrict behaviour, and the second is a measure of the degree of effort applied to attempting to control pain. Taken together these represent the overall process of accep-tance of chronic pain. The scale overall internal consistency is 0.78. *Acceptance and action questionnaire—II* (AAQ-II) (Bond et al., 2011). This is a measure of psychological inflexibility/acceptance. It is a seven item scale producing a single score, ranging from seven to forty-nine; higher scores suggest lower levels of acceptance or willingness to remain in contact with unpleasant private experi-ences. Its internal consistency is 0.84.

Intervention

Treatment was adapted for use with an individual from the group protocol used by Tooze and colleagues (unpublished), which itself was based on the principles and methods developed by Steven C. Hayes and colleagues (Hayes et al., 2012). The six main treatment processes utilised in ACT match those represented Figure 1 and were designed to address the formulation of W's distress based on that model. It involved six one-hour sessions of individual treatment with each session corresponding to one of the processes in the "hexaflex". A mainstay of ACT treatment is the use of metaphor to help undermine patients' "language induced struggle . . . without invoking the client's normal verbal defenses" (Hayes et al., 2003, p. 76) and was used throughout treatment. Details of the specific exercises used in the sessions are provided in most of the treatment manuals based on ACT (Hayes et al., 2012; McCracken, 2006; McCracken et al., 2004). Exercises that had been planned for the session but had not been completed were given to W as homework assignments. For example, at the end of session two, W was given an exercise called "filling the head", which was a worksheet with an outline of a human head. W was asked to write on the sheet any troublesome emotions, memories, thoughts, sensations, or urges and then to consider these as "fellow travellers" with her initial pain. She was asked if she would be willing to carry this metaphorical head in her pocket for a while.

Sessions One and Two

Treatment began by exploring what W had been doing to manage her pain so far. Her main attempts to manage her pain had been through medication and by changes to activities during the day such as resting in the early evening. The "Tug of war with a monster" and "Man in a hole" metaphor were used to help W begin to accept that her attempts at control had not been successful and had actually exacerbated her distress. The aim of the session was to help W begin to build a willingness to contact feelings of pain and emotional distress without trying to block or control them (the "Chinese handcuffs" metaphors (see Appendix 2) was used to explore how to have a different relationship with distress, that is, struggling against the handcuffs only makes them tighten) and to engage in valued activities even in their presence (the "Passengers on the bus" was used to help explore this (see Appendix 2)).

Session Three

In the third session cognitive de-fusion entailed the use of *in vivo* exercises to explore how thinking can have an unhelpful influence on behaviour and restrict and distort the perception of what is really happening in the patient's life. The goal is to loosen the grip of thinking so that the person can respond to events in a more flexible way. Here the "Milk" exercise was used. W was asked to say the word "milk" and notice what comes to mind (usually a glass of milk and the cold, creamy texture when drinking it). W was then asked to join the therapist in repeating the word out loud over and over as rapidly as possible for thirty seconds. The exercise highlights how the word "milk" has become fused with our experience of it, but by repeating it over and over, the word can be "de-fused" from experience and seen for what it actually is: a sound produced by the vocal cords. This helps the patient see that words and language do not have to be seen as literal truth.

Session Four

Mindfulness techniques such as meditation were used in the fourth session to help W become more aware of the present moment, as well as sensations in the body. W began to notice that her pain changed from moment to moment and was not static, even at times subsiding all together even without medication. Meditation practise also helped W see that she was spending a lot of her time caught up in ruminating about the past or worrying about the future, rather than paying attention to what was happening for her in the present moment. She noted that this practise helped her to appreciate more what was happening in her life right now.

Session Five

The session on "self as context" attempted to build awareness in W that her thoughts do not constitute the sum total of who she is and developing the ability to observe thoughts and build distance between herself and her thoughts. An exercise that was used here was for W to consider some of the labels she attributes to herself, such as answers to the following; "I am a person who . . .", "I am a person who cannot . . .", "the worst thing about me is . . .". She was then asked to consider any investment she had in these labels, especially the negative ones and whether she defends these as if the labels were true, rather than thoughts about herself. W noted that she had become very attached to her pain and it was a revelation to her that she could continue her life without having to pay so much attention to it.

Session Six

The final part of treatment consists of clarifying values and building commitment to behave in ways that move them towards their values even in the face of pain or failure. W was asked to consider one value that was important to her and what action she could take, however small, to move towards that value. Building commitment to act involved planning to behave in ways consistent with her values by setting goals that helped her move in their direction.

Adaptations to treatment

W has stated that her main complaint was pain and its impact was worse later in the day. As a result we organised sessions for early in the morning. While she did not think fatigue was an issue, it did become evident during sessions. Hence additional time was built into sessions and written material on the session topic was provided at the end of each session. Worksheets were provided for W to complete and any potential obstacles to completing the homework were identified and problem solving was used to overcome them. See Table 1 for a summary of the treatment sessions.

Analysis

Visual inspection of the data was used to assess the change in measures across treatment and at follow-up. Reliable change in scores was calculated using the method developed by Jacobson and Truax (1991) whereby the difference between the pre- and post-treatment scores on the measures are divided by the standard error of the difference between the two test scores (see Appendix 1). To calculate this index, the pre-treatment standard deviations of the measures reported in the study by Tooze and colleagues (unpublished) were used as representative of MS patients suffering chronic neuropathic pain. An RCI above 1.96 suggests that the change in score is statistically significant ($\sim = 0.05$). To assess whether change in W's scores from pre-treatment were clinically significant, Criteria A from Jacobson and Truax (1991) was used. That is, where there has been a reliable change, the post-treatment score is more than 1.96 standard deviations from the clinical group sample mean (in this case from the Tooze et al. (unpublished) sample).

Outcomes

W had entered therapy seeking a mindfulness-based approach and she engaged well with the mindfulness exercises and observing her

Table 1: Summary of treatment sessions including techniques used and the ACT treatment process they relate to

Session	ACT process	Intervention
1. Undermining experiential control	Acceptance	Exploring attempts to control/manage pain. *Metaphors: "Man in a hole", "Tug of war with a monster"*
2. Building willingness skills	Acceptance	Develop willingness to act in linewith values in presence of pain. *Metaphors: "Passengers on the bus", "Chinese handcuffs"*
3. Cognitive de-fusion	Cognitive de-fusion	De-fusion of the literality of words: *"Milk" exercise*
4. Contact with present moment	Contact with present moment	Mindfulness meditation: *Meditation on breath and body*
5. Observer self	Self-as-context	Mindfulness (observer self): *Observing thoughts* The conceptualised self: *Exploring the labels we give ourselves*
6. Values clarification and committed action	Values, Committed action	Psycho-education: *Values clarification exercise and goal setting*

thoughts. She found the famous "Passengers on the bus" metaphor from ACT particularly useful and helped her to see that she could continue with her life without having to control every element. More difficult were the exercises around self-as-context and eliciting values. W felt she was very clear what her values were, but acknowledged towards the end of therapy that she had perhaps not engaged in valued actions to the extent she previously thought. This was highlighted by a trip to the cinema when she realised she had sat through a two-hour film without trying to control her pain but just accepted that she was in pain.

Table 2 shows that at baseline assessment, W's score on the PES was twenty-nine, close to the maximum score of thirty, suggesting that pain was impacting on her functioning over the previous four weeks. This is in line with her pre-treatment rating of pain intensity (BPI), which was nine (with ten being the maximum). Although fatigue was not a main complaint for W, her score on the MFIS was in the mid-range suggesting that it was impacting on her daily activities. However, while pain and fatigue were impacting on functioning, her QLI score of 23.91 suggests this was not related to a low perceived quality of life.

Process measures of change during treatment are shown in Figure 2. At the start of treatment, W's CPAQ score was sixty-six and rose steadily across treatment to 116 with the increase remaining stable at two month follow-up. The statistical significance of the improvement is shown by the RCI score of 4.33 ($p < 0.05$) and was also a clinically significant change. Scores on the AAQ-II decreased steadily across treatment sessions from nineteen to seven at the end of treatment and remained the same at follow-up. The RCI score for this change was not statistically significant (RCI = 1.59).

Based on the RCIs in Table 2, at the end of treatment W showed statistically significant decreases on measures of the impact of pain (PES) and impact of fatigue (MFIS-5), which were maintained at follow-up ($p < 0.05$). These changes were also clinically significant improvements. Her pain intensity score at post treatment also showed significant change based on its RCI and improved further at follow-up ($p < 0.05$) with the change being clinically significant. Her subjective ratings of quality of life (QLI) did show improvement across pre-, post-treatment, and follow-up but was not statistically significant.

Table 2: Outcome measures at pre- and post-treatment and follow up

Measure	Pre-Treatment	Post-Treatment	Follow-up	RCI
BPI	9	5	3*	6.30
PES	29	11	9*	7.72
MFIS-5	9	3	2*	2.70
QLI	23.91	25.36	26.49	1.01

Note: BPI = Brief Pain Inventory; PES = Pain effects scale; MFIS-5 = Modified Fatigue Impact Scale; QLI = Quality of Life Index; * = Clinically significant change.

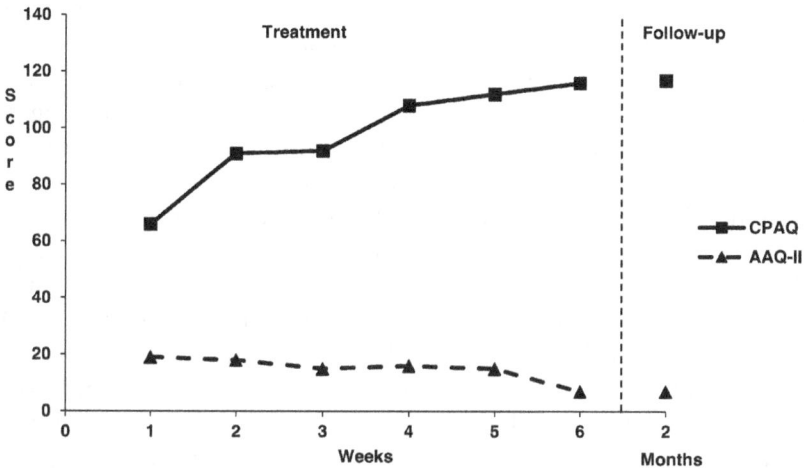

Figure 2: Process measures across treatment sessions.

Note: CPAQ = Chronic Pain Acceptance Questionnaire; AAQ-II = Acceptance and Action Questionnaire.

Discussion

Over the course of treatment W experienced significant improvement on measures of the impact of pain (PES) and fatigue (MFIS-5) on her life as well as the intensity of her pain (BPI). At post-treatment and follow-up her subjective quality of life (QLI) had risen slightly and her level of psychological inflexibility fell (albeit not statistically significant). These findings were as expected given previous research that found ACT led to significant improvement on pain disability/behaviour and life satisfaction/QoL (Sheppard et al., 2010; Wicksell et al., 2008).

The ACT literature suggests the reason for improvement in measures of QoL and impact of pain is not due to changes in pain symptoms per se, but on acceptance of pain and to "act effectively in accordance with personal values in [its] presence" (McCracken & Gutiérrez-Martínez, 2011; McCracken et al., 2004; Wicksell et al., 2010, p. 1059). This appeared to be the case with W. Her initial CPAQ score of sixty-six was similar to mean scores found in pain patients in previous research (e.g., McCraken et al., 2004) and rose significantly over treatment and remained stable at follow-up. W's AAQ-II scores (a measure of general psychological inflexibility) fell over the course of treatment, from nineteen to seven, but the change was not significant. Furthermore, W's initial score of nineteen on this measure would indicate that W did not fall within the clinical range seen in other samples (see Bond et al., 2011) and may suggest that general psychological inflexibility was not a significant factor maintaining W's distress. The finding of a significant change in CPAQ score but not in AAQ-II is in line with the findings of Tooze and colleagues (unpublished) who used the same ACT protocol for a group treatment of pain sufferers with MS. These authors explained the finding with reference to a regression analysis conducted by McCraken and Zhao-O'Brien (2010), which suggested that general and pain specific acceptance made unique contributions to predicting patient functioning and were empirically distinguishable. Given that acceptance of pain can be conceptualised as a sub-set of behaviours falling within the larger set of general acceptance, the brief nature and specific focus of the intervention on pain within MS may not have led to significant changes in general level of acceptance/flexibility to be detected by the AAQ-II. It would also suggest that the main mechanism of change over the course of treatment was W's acceptance of pain rather than her psychological flexibility per se; which is supported by her self-report during therapy (see below). It could be argued that the significant reduction in pain intensity (BPI) might account for the improvements in outcome measures. Previous research indicates that a goal of ACT in chronic pain is not necessarily a reduction in pain symptoms (see Wicksell et al., 2010). W's self-report was that her pain had not changed over therapy so it is not clear whether her response to the BPI represents an accurate assessment of her experience of pain.

A crucial point in treatment was when W realised how attached she was to her pain and how important it had become in her life. After the third session, W began to realise that she could not control her pain, and this corresponded to further increases in her CPAQ score. She felt this was due to her letting go of her attachment to pain and accepting that it was there, as opposed to her old ways of trying to alleviate it. At end of therapy in anecdotal comments, W said she felt she had moved from a position of trying to control her pain to one where she accepted that she cannot control it yet can still move on with her life. She acknowledged that this acceptance of her situation also helped her manage her other MS related symptoms such as fatigue. Initially W had not seen this as problematic but over the course of treatment she realised that she was starting to give herself permission to rest; whereas previously she would have felt guilt when, for example, she could not engage in housework. At the end of therapy W was less worried about falling, and she saw this as due to being less anxious and focused on her condition. Overall, she said she felt the change was "extraordinary" and had been surprised how quickly it had happened. This acceptance of a lack of control over the symptoms of MS following an ACT intervention is similar to the case reported by Gillanders and Gillanders (2014) whose MS patient also experienced a lack of control over her MS but came to a realisation following treatment that she was able to "live with the condition" (p. 36). The course of treatment followed closely the topic guide developed for the group treatment by Tooze and colleagues (unpublished) with each individual session covering the topics for each group session. As a result it was not as idiosyncratic as it could have been if following an individualised case conceptualisation. Topics were covered that perhaps may not have been addressed through a more individualised approach. One such example was the topic of Session Six: values. W was quite clear about her values and was able to express them cogently. As mentioned above, in comments she made at the end of therapy, she felt that it was her acceptance of her lack of control over her pain that allowed her to engage in valued behaviours. Hence from this particular case, it is unclear whether all areas of the hexaflex need addressing during an ACT intervention. For W, acceptance of her lack of control over pain appeared central to her increase in psychological flexibility, clarifying her values less so. Given that issues of self-identity, vulnerability, and control are common themes seen in MS (Gillanders & Gillanders, 2014) it seems appropriate that these are addressed in psychological interventions for this population. It is unclear whether a treatment that followed a more individualised case conceptualisation would have led to even better outcomes.

This is important because W was fortunate to have retained a relatively high level of physical functioning (given a twenty year history of MS) and she was also very motivated and seeking a mindfulness based intervention. Hence W was very receptive to the concepts of ACT discussed during sessions and is likely to have led to the dramatic improvement in measures of acceptance (e.g., CPAQ) and her anecdotal comments at the end of therapy. Would a patient suffering greater levels of physical and cognitive impairment, and less affinity with mindfulness based approaches, have improved to the extent W did from the generic ACT for pain approach presented here? Further research on the feasibility of ACT for pain in MS patients will help to clarify this but the findings of Tooze and colleagues (unpublished) are encouraging and suggest this is a promising avenue to explore. While it could be argued that improvements in psychological flexibility are unlikely given an underlying neurological impairment, we concur with Gillanders and Gillanders) that ACT may help mitigate the effects of neuropsychological problems through its behavioural focus "designed to help people slow down, increase awareness, take perspective, and change habitual modes of responding" (p. 37). However, as yet this remains an empirical question for further research.

One of the main limitations of the present study is the lack of multiple baselines to ensure stability of the measures prior to treatment, ensuring any change is attributable to therapy (Barlow & Hersen, 1984; Hayes, 1981). Given this, Tate and colleagues (2013) state that the study design used here would not meet criteria for a single case experimental design, making it difficult to attribute the effects reported to the therapy intervention. Recent developments have seen the inclusion of randomised n-of-1 trials as Level 1 evidence for treatment decisions according to Oxford Centre for Evidence-Based Medicine. This led Tate and colleagues (2013) to develop the risk of bias in n-of-1 trials (RoBiNT) Scale to assess the quality of single case experimental designs and to help researchers to develop rigorous single case experimental designs. Utilising this scale and the guidelines in future research with single patients would help to develop a robust evidence base for using ACT in MS patients.

Even with the potential limitations in the present study's design, given that W's difficulties with controlling pain had persisted for a number of years, there is no reason to think that any of the measures used would not have been stable over a prolonged baseline. A further limitation of a single case report is the generalisability of the findings. Tooze and colleagues (unpublished) used a similar treatment protocol in a group format and found similar significant change in the level of pain acceptance between pre- and post-treatment. With individual sessions over a longer treatment period (six as opposed to four), the current study found significant improvements in acceptance of pain, impact of pain and fatigue, and pain intensity. Given the similar level of training and experience of the therapist, in this case in using

ACT, it would suggest that individual treatment sessions may be of more benefit. However, W had specifically sought out a "mindfulness" based treatment and may have been more engaged with its approach than the group members in the Tooze and colleagues (unpublished) study. Further research with larger sample sizes is needed to understand whether ACT is generally effective in this patient population and to confirm whether the main mechanism of change is acceptance of pain, rather than more general psychological flexibility, as suggested by the finding of this study and Tooze and colleagues.

To our knowledge this is the first case report of ACT being used for the treatment of chronic neuropathic pain in MS. The treatment lead to significant improvement on a number of treatment measures and the patient reported that the quality of her life improved. The study suggests that ACT is a viable alternative to treating chronic pain in this patient group and can provide the patient a choice of treatments when standard CBT approaches are not deemed acceptable.

Appendix 1—Reliable Change Index (Jacobson & Traux, 1991)

Let X_1 = Pre-test score; X_2 = Post-test Score; r_{xx} = Test-retest reliability of measure; S_{diff} = standard error of difference between the two test scores; S_E = standard error of measurement;

Reliable change index (RC) is calculated by the following equation:

$$RC = \frac{X_2 - X_1}{S_{diff}}$$

where
$$S_{diff} = \sqrt{2}(S_E)^2$$

and
$$S_E = S_x \sqrt{(1 - r_{xx})}$$

An RC larger than 1.96 would be unlikely to occur (p < 0.05) without actual change. When RC exceeds this level the individual can be classified as reliably changed.

Appendix 2

Metaphor: passengers on the bus

It is like you are a bus driver and you want to go where you want to go. At the same time on this bus are these scary passengers. They don't always want to go where you want to go, and when you don't go their way they let you know about it. They may rush up behind you, crawl all over you, and threaten you. They essentially bully you so you do what they say. You choose not to go where you want to go and they settle down, into the back of the bus and out of sight. In the meantime you're driving around in circles and not going anywhere in particular, just driving aimlessly. Now you may get fed up with this eventually. You may stop the bus and try to toss these passengers off, but there are many and they fight you. And notice that all the time you fight them the bus is not going anywhere. And so it's back to the old agreement, if they leave you alone you will only go where they say and nowhere else. Notice this interesting part, the key thing, these passengers have never done you any physical harm, they cannot, and never will. All they got over you is the ability to intimidate. The only power they have over you is the power you give them. You are the driver yet you trade your control over the bus to keep the passengers away. You may say that this is silly or that you do not have to put up with this. The truth is you do have passengers and they are your thoughts, feelings, sensations, urges, memories, and the like.

Metaphor: Chinese handcuffs

I wonder if the situation here is something like this (hand patient Chinese finger trap).

Did you ever play with these when you were a kid? We called them "Chinese handcuffs". They are also called Chinese finger traps. Check this out. This is just a tube of woven straw. Now, push both index fingers in, one into each end, and see what happens. You notice that as you pull them back out, the straw catches and tightens. You may notice other things that happen, such as in your feelings or thoughts. What's happening here? See, the harder you pull, the smaller the tube gets and the tighter it holds your fingers.

Maybe this situation with pain, distress, and the other experiences come with it is something like this trap. Maybe there is no healthy way to get out of pain or distress once we are stuck in it, such as when it is a chronic condition, and any attempt to do so just restricts your room to move. Have you noticed something else about this little tube?

With this little tube, the only way to get some room is to push your fingers in, which makes the tube bigger. That may be hard to do at first, because everything your mind tells you casts the issue in terms of "in and out" not "tight and loose". But your experience is telling you that if the issue is "in and out", then things will be tight. Maybe you need to come at this situation from a whole different angle, different than what your mind tells you to do with your experience of suffering.

Is this "moving in" something they could do when they are struggling to get out of experiences outside of session? Let's identify some possible situations.

References

American Psychological Association (2014). Chronic or persistent pain in general (including numerous conditions). Available at: www.div12. org/psychological-treatments/disorders/chronic-or-persistent-pain/ (last accessed 2 March 2014).

Barlow, D. H., & Hersen, M. (1984). *Single Case Experimental Designs: Strategies for Studying Behavior Change* (2nd edn). New York: Pergamon.

Beniczky, S., Tajti, J., Varga, E. T., & Vecsei, L. (2005). Evidence-based pharmacological treatment of neuropathic pain syndromes. *Journal of Neural Transmission, 112*(6): 735–749.

Bond, F. W., Hayes, S. C., Baer, R. A., Carpenter, K. M., Guenole, N., Orcutt, H. K., Waltz, T., & Zettle, R. D. (2011). Preliminary psychometric properties of the Acceptance and Action Questionnaire–II: a revised measure of psychological inflexibility and experiential avoidance. *Behavior Therapy, 42*(4): 676–688.

Ferrans, C. E., & Powers, M. J. (2007). Psychometric assessment of the Quality of Life Index. *Research in Nursing & Health, 15*(1): 29–38.

Foley, P. L., Vesterinen, H. M., Laird, B. J., Sena, E. S., Colvin, L. A., Chandran, S., MacLeod, M. R., & Fallon, M. T. (2013). Prevalence and natural history of pain in adults with multiple sclerosis: systematic review and meta-analysis. *Pain, 154*(5): 632–642.

Forbes, A., While, A., Mathes, L., & Griffiths, P. (2006). Health problems and health-related quality of life in people with multiple sclerosis. *Clinical Rehabilitation, 20*(1): 67–78.

Gillanders, S., & Gillanders, D. (2014). An acceptance and commitment therapy intervention for a woman with secondary progressive multiple sclerosis and a history of childhood trauma. *Neuro-Disability and Psychotherapy, 2*(1/2): 19–40.

Graham, C. D., Gillanders, D., Stuart, S., & Gouick, J. (2015). An Acceptance and Commitment Therapy (ACT)–based intervention for an adult experiencing post-stroke anxiety and medically unexplained symptoms. *Clinical Case Studies, 14*(2): 83–97.

Harrison, S. L., Robertson, N., Graham, C. D., Williams, J., Steiner, M. C., Morgan, M. D. L., & Singh, S. J. (2014). Can we identify patients with different illness schema following an acute exacerbation of COPD: a cluster analysis. *Respiratory Medicine, 108*: 319–328.

Hawthorne, G., Richardson, J., & Osborne, R. (1999). The Assessment of Quality of Life (AQoL) instrument: a psychometric measure of health-related quality of life. *Quality of Life Research, 8*(3): 209–224.

Hayes, S. C. (1981). Single case experimental design and empirical clinical practice. *Journal of Consulting and Clinical Psychology, 49*(2): 193–211.

Hayes, S. C., & Duckworth, M. P. (2006). Acceptance and commitment therapy and traditional cognitive behavior therapy approaches to pain. *Cognitive and Behavioral Practice, 13*(3): 185–187.

Hayes, S. C., Barnes-Holmes, D., & Roche, B. (2001). *Relational Frame Theory: A Post-Skinnerian Account of Human Language and Cognition.* New York: Springer.

Hayes, S. C., Luoma, J. B., Bond, F. W., Masuda, A., & Lillis, J. (2006). Acceptance and commitment therapy: model, processes and outcomes. *Behaviour Research and Therapy, 44*(1): 1–25.

Hayes, S. C., Masuda, A., & De Mey, H. (2003). Acceptance and commitment therapy and the third wave of behavior therapy (Acceptance and commitment therapy: een derde-generatie gedragstherapie). *Gedragstherapie (Dutch Journal of Behavior Therapy),* 2: 69–96.

Hayes, S. C., Strosahl, K. D., & Wilson, K. G. (2012). *Acceptance and Commitment Therapy: The Process and Practice of Mindful Change* (2nd edn). New York: Guilford Press.

Heckman-Stone, C., & Stone, C. (2001). Pain management techniques used by patients with multiple sclerosis. *The Journal of Pain, 2*(4): 205–208.

IASP (2007). International Association for the Study of Pain Taxonomy. Available at: http://www.iasp-pain.org/Taxonomy (last accessed 26 February 2014).

Jacobson, N. S., & Truax, P. (1991). Clinical significance: a statistical approach to defining meaningful change in psychotherapy research. *Journal of Consulting and Clinical Psychology, 59*(1): 12–19.

Kangas, M., & MacDonald, S. (2011). Is it time to act? The potential of acceptance and commitment therapy for psychological problems following acquired brain injury. *Neuropsychological Rehabilitation: An International Journal, 21*(2): 250–276.

Khan, F., & Pallant, J. (2007). Chronic pain in multiple sclerosis: prevalence, characteristics, and impact on quality of life in an Australian community cohort. *Journal of Pain, 8*(8): 614–623.

McCracken, L. M. (2006). *Contextual Cognitive-Behavioral Therapy for Chronic Pain.* Seattle, WA: International Association for the Study of Pain.

McCracken, L. M., & Gutiérrez-Martínez, O. (2011). Processes of change in psychological flexibility in an interdisciplinary group-based treatment for chronic pain based on acceptance and commitment therapy. *Behaviour Research and Therapy, 49*(4): 267–274.

McCracken, L. M., & Zhao-O'Brien, J. (2010). General psychological acceptance and chronic pain: there is more to accept than the pain itself. *European Journal of Pain, 14*: 170–175.

McCracken, L. M., Vowles, K. E., & Eccleston, C. (2004). Acceptance of chronic pain: component analysis and a revised assessment method. *Pain, 107*(1): 159–166.

Morley, S., Williams, A., & Eccleston, C. (2013). Examining the evidence about psychological treatments for chronic pain: Time for a paradigm shift? *Pain, 154*(10): 1929–1931.

Moulin, D. E., Foley, K. M., & Ebers, G. C. (1988). Pain syndromes in multiple sclerosis. *Neurology, 38*(12): 1830–1830.

NICE (2013). Neuropathic pain—pharmacological management: the pharmacological management of neuropathic pain in adults in non-specialist settings. NICE clinical guideline 173. Available at www.nice.org.uk/guidance/cg173 (last accessed 3 August 2015).

O'Connor, A. B., Schwid, S. R., Herrmann, D. N., Markman, J. D., & Dworkin, R. H. (2008). Pain associated with multiple sclerosis: systematic review and proposed classification. *Pain, 137*(1): 96–111.

Reading, A. E., Everitt, B., & Sledmere, C. M. (1982). The McGill Pain Questionnaire: a replication of its construction. *British Journal of Clinical Psychology, 21*(4): 339–349.

Richards, R., Simpson, F., Beard, S., & Tappenden, P. (2002). A review of the natural history and epidemiology of multiple sclerosis: implications for resource allocation and health economic models. *Health Technology Assessment, 6*(10): 1–73.

Ritvo, P., Fischer, J., Miller, D., Andrews, H., Paty, D., & LaRocca, N. (1997). *Multiple Sclerosis Quality of Life Inventory: A User's Manual.* New York: National Multiple Sclerosis Society.

Rossi, S., Mataluni, G., Codeca, C., Fiore, S., Buttari, F., Musella, A., Castelli, M., Bernardi, G., & Centonze, D. (2009). Effects of levetiracetam on chronic pain in multiple sclerosis: results of a pilot, randomized, placebo controlled study. *European Journal of Neurology*, *16*(3): 360–366.

Sheppard, S. C., Forsyth, J. P., Hickling, E. J., & Bianchi, J. (2010). A novel application of acceptance and commitment therapy for psychosocial problems associated with multiple sclerosis: results from a half-day workshop intervention. *International Journal of MS Care*, *12*(4): 200–206.

Sherbourne, C. D. (1992). Pain measures. In: A. L. Stewart & J. E. Ware (Eds.), *Measuring Functioning and Well-being: the Medical Outcomes Study Approach* (pp. 220–234). London: Duke University Press.

Solaro, C., & Uccelli, M. M. (2011). Management of pain in multiple sclerosis: a pharmacological approach. *Nature Reviews Neurology*, *7*(9): 519–527.

Stuifbergen, A. K. (1995). Health-promoting behaviors and quality of life among individuals with multiple sclerosis. *Research and Theory for Nursing Practice*, *9*(1): 31–50.

Sylvester, M. (2011). Acceptance and commitment therapy for improving adaptive functioning in persons with a history of pediatric acquired brain injury. Dissertation Abstract, University of Nevada, Reno.

Tate, R. L., Perdices, M., Rosenkoetter, U., Wakim, D., Godbee, K., Togher, L., & McDonald, S. (2013). Revision of a method quality rating scale for single-case experimental designs and n-of-1 trials: the 15-item Risk of Bias in N-of-1 Trials (RoBiNT) Scale. *Neuropsychological Rehabilitation*, *23*(5): 619–638.

Tooze, O. J., Karl, A., Dysch, L., & McLaughlin, D. (2014). An exploration of acceptance and commitment therapy for chronic pain in multiple sclerosis. *Neuro-Disability and Psychotherapy*, *2*(1): 1–18.

Tooze, O. J., Karl, A., McCracken, L. M., & Dysch, L. (unpublished). Acceptance and Commitment Therapy for chronic pain with people who have multiple sclerosis: a feasibility study. Unpublished manuscript.

Turk, D. C., Rudy, T. E., & Salovey, P. (1985). The McGill Pain Questionnaire reconsidered: confirming the factor structure and examining appropriate uses. *Pain*, *21*(4): 385–397.

Wetherell, J. L., Afari, N., Rutledge, T., Sorrell, J. T., Stoddard, J. A., Petkus, A. J., Soloman, B. C., Lehman, D. H., Liu, L., Lang, A. J., & Atkinson, J. H. (2011). A randomized, controlled trial of acceptance and commitment therapy and cognitive-behavioral therapy for chronic pain. *Pain*, *152*(9): 2098–2107.

Whiting, D. L., Simpson, G. K., Mcleod, H. J., Deane, F. P., & Ciarrochi, J. (2012). Acceptance and commitment therapy (ACT) for psychological adjustment after traumatic brain injury: reporting the protocol for a randomised controlled trial. *Brain Impairment*, *13*(3): 360–376.

Wicksell, R. K., Ahlqvist, J., Bring, A., Melin, L., & Olsson, G. L. (2008). Can exposure and acceptance strategies improve functioning and life satisfaction in people with chronic pain and whiplash-associated disorders (WAD)? A randomized controlled trial. *Cognitive Behaviour Therapy*, *37*(3): 169–182.

Wicksell, R. K., Olsson, G. L., & Hayes, S. C. (2010). Psychological flexibility as a mediator of improvement in Acceptance and Commitment Therapy for patients with chronic pain following whiplash. *European Journal of Pain*, *14*(10): e1051–e1059.

Williams, A. C., Eccleston, C., & Morley, S. (2012). Psychological therapies for the management of chronic pain (excluding headache) in adults. *Cochrane Database Systematic Review, 11*.

CHAPTER THREE

An acceptance and commitment therapy intervention for a woman with secondary progressive multiple sclerosis and a history of childhood trauma*

Sarah Gillanders and David Gillanders

Theoretical and research basis

Multiple sclerosis (MS) is a chronic neurological condition characterised by demyelination and atrophy within the central nervous system. Secondary progressive multiple sclerosis (SPMS) occurs following relapsing remitting multiple sclerosis, and is characterised by a reduction or cessation in the number of relapses, and a gradual increase in disability (Vukusic & Confavreux, 2003). As with other types of MS, SPMS is idiosyncratic. For some people the progression is gradual, but for others it is rapid. Interventions become focused on symptom management, preventing complications, and promoting health and well-being, in the context of ongoing decline of function. While some people find this stage of the disease more predictable and easier to live with, many people find it psychologically challenging due to the severity of symptoms and their impact on quality of life. Jones and colleagues (2012) studied the responses of 4178 people with MS on the UK MS register and found that people with SPMS were significantly more likely to be depressed than those with other types of MS.

* Originally published in 2014 in *Neuro-Disability & Psychotherapy*, 2(1/2): 19–40.

While there has been little research to date on the impact of child-hood trauma on well-being in adults with MS, these relationships have been investigated in other medical conditions. For example, Coker and colleagues (2012) studied 553 women with breast, cervical, or colorectal cancer and found that childhood sexual abuse may affect depression, perceived stress, and cancer-related well-being. Roy and colleagues (2011) examined the relationship between childhood trauma and depressive symptoms in 150 African Americans with type one diabetes and found that patients with Beck Depression Inventory (BDI) scores above fourteen had experienced significantly more types of childhood trauma than those with BDI scores below this cut off. Despite the established association between childhood abuse and poorer well-being in chronic diseases, one study found that some survivors used strategies they had learned from past experience to cope adaptively with the treatment-related side effects of hepatitis C (Hopwood & Treloar, 2008).

A mediating factor between early adversity and living successfully with MS is resilience. Resilient coping is defined as the "ability to maintain relatively stable and healthy levels of psychological and physical functioning when confronted with a highly disruptive situation" (Bonanno, 2004, p. 20). This type of resilience has an emphasis on chronic and uncontrollable stressors, and the social-contextual factors that promote adaptation. MS is just such a chronic, unpredictable condition that impacts significantly on social functioning and relationships (Mutch, 2010). Incorporating information about life history and responses to adversity into clinical assessment with people with MS can help clinicians to understand both the factors that make people vulnerable to distress and the signs of resilience that can be used to support well-being.

To date, the best-evidenced approaches to psychological intervention in MS have been based on cognitive behavioural therapy (CBT) (e.g., Moss-Morris et al., 2010, 2013). CBT approaches stem from a conceptualisation of human suffering being based upon the meaning we make of our lives. In psychological disorders, the CBT model suggests that individuals interpret and appraise symptoms, situations, and their own responses to these. Critically, in the cognitive model, a distinct pattern of cognitive biases is postulated to influence our emotions and behaviours (Alford & Beck, 1997). In the CBT model of psychopathology, such biases or "cognitive distortions" result in heightened emotional and behavioural responses, leading to states of psychological disorder such as anxiety and depression. The core focus in a CBT approach is towards identification of maladaptive emotional responses, the underpinning cognitive distortions that are hypothesised to be influencing these and the maladaptive behavioural responses that are postulated to be driven by and maintaining these emotional and cognitive patterns (Beck, 2011).

In MS for instance, a cognitive approach might look for themes within thinking that suggest the person is appraising the future with MS as "catastrophic" (i.e., the worst possible outcome). Such thoughts would be considered to be an underpinning cognitive bias that drives emotions such as anxiousness. On the other hand, cognitive appraisals of loss or personal helplessness would be more likely to lead to emotions such as sadness, and appraisals of the diagnosis of MS as unfair would more likely lead to anger. In traditional CBT approaches, the therapist would take the stance of helping to educate the patient to become his or her own therapist. The CBT relationship is highly collaborative and the patient is invited to adopt an empirical stance to their problem. This stance involves a search for these characteristic "cognitive distortions", evaluation of the evidence the person is using to support such appraisals, generation of alternative appraisals, and reality testing through behavioural experiments. The underlying rationale of these strategies is to change distorted cognitive representations of self, world, symptoms, and others, in order to reduce maladaptive emotional and behavioural responses (see e.g., Beck, 2011).

Existing CBT approaches to MS combine reappraisal of the meaning of symptoms and modification of thoughts about MS with other strategies such as acceptance, goal setting, and symptom management. While traditional CBT approaches do explicitly target acceptance of MS, recent developments in CBT place a far greater emphasis on acceptance. In both of these modalities, acceptance is presented as an alternative to avoidance coping.

Avoidance based strategies take many forms, though share the function of trying to not be in contact with unwanted private events, such as symptoms, thoughts, and emotions. Examples include not talking about MS, neglecting health care needs, social withdrawal, and not using potentially useful aids and adaptations. This can result in great costs to the individual and family, such as reduced opportunities for intimacy, loss of friendship, relationship strain, a sense of isolation, as well as deconditioning and disability.

Acceptance and commitment therapy is a form of cognitive behavioural therapy that is rooted in a behaviour analytic conceptualisation of the links between cognition, affect, and behaviour (Hayes et al., 2012). Instead of a cognitive model of psychopathology, ACT rests upon a behavioural model of human functioning that encompasses states that we think of as "disordered" as well as "normal function". ACT takes a non-eliminative approach; rather than developing strategies to eliminate symptoms or modifying the content of thoughts, ACT teaches acceptance and mindfulness strategies to help people become more willing to experience unpleasant sensations, feelings, and thoughts. At the same time, the client is helped to explore and contact what they most deeply care about in their life; their values. Values are freely chosen, abstract principles that specify what is important in a person's life, such as being a loving parent, being a committed life partner, or being of service to a community.

Rather than persisting with ineffective attempts to control or avoid unpleasant internal experiences, clients in ACT learn to let go of struggling, developing greater willingness to have such experiences. In doing so, greater choice of behaviour is afforded, allowing life to become about pursuing values-consistent goals, even in the presence of anxiety, fatigue, and uncertainty.

While in traditional CBT approaches clients are taught how to monitor and change their thinking to become less catastrophic and more balanced, in an ACT approach there is a greater recognition that thoughts about MS may be realistic. Under these circumstances, clients are taught skills to step back from thoughts and take a distanced perspective on distressing content, even if that content is "true". ACT therapists use experiential exercises, such as imagining one's thoughts as "passengers on a bus" or as if thoughts are leaves on a stream. Such exercises use guided and personally relevant imagery to facilitate the patient having the experience of stepping back and seeing difficult thoughts "as if" they were outside of and separate from the self. In addition, ACT therapists use language in playful ways in order to point at thought processes, so that the effects of thinking on behaviour can be seen more clearly and responded to with greater choice. ACT also uses metaphorical talk to highlight and influence such relationships between thinking and acting, but without the explicit goal to try and change the content of beliefs. Rather, the aim is to undermine the power of cognitive events to lead automatically to behavioural responses (Blackledge, 2007).

For example, rather than becoming entangled in thoughts about the eventual progression of MS; such naturally distressing content is first acknowledged and is related to in a compassionate manner. An ACT therapist might ask the client to imagine what these thoughts of progression would look like if they could be seen outside of the self. By describing and experiencing these thoughts in this way, with a present focus, but from a perspective of looking at these thoughts, rather than from these thoughts, the patient begins to experience a sense of distance from even very difficult or painful thoughts. In the context of a supportive, validating, equal, and courageous therapeutic relationship, even difficult thoughts such as these might be responded to in ways that undermine their impact, using humour, observation, mindfulness, and perspective taking. The patient learns through experience that over time, rather than getting into struggles to avoid such thoughts, they can be present, without needing to be controlled, resolved, or responded to behaviourally. Unlike the cognitive model, at no point is the patient encouraged to think differently, think more adaptive or more realistic thoughts, to change the content, frequency, or form of their thoughts as a route to reducing any symptoms or emotional reactions. In essence; life becomes gradually more about pursuing what matters, and distressing thoughts and feelings are responded to compassionately, welcomed as part of being human, rather than being things that need to be fixed before living can happen. The ability to be fully present with distressing private events and still persist towards valued goals is known as psychological flexibility.

The ACT model (see Figure 1) describes six overlapping and interdependent behavioural processes that lead to inflexibility (left hand side) and the corresponding processes that therapists can use to develop client psychological flexibility (right hand side). These flexibility processes are: willingness (choosing to make room for difficult internal events, letting go of needless defence or attempts to get rid of them), contact with the present moment (paying attention to the here and now, rather than being caught up in the future or past), cognitive defusion (stepping back from thoughts, seeing them less literally and more as mental events, even if they are true), contact with values

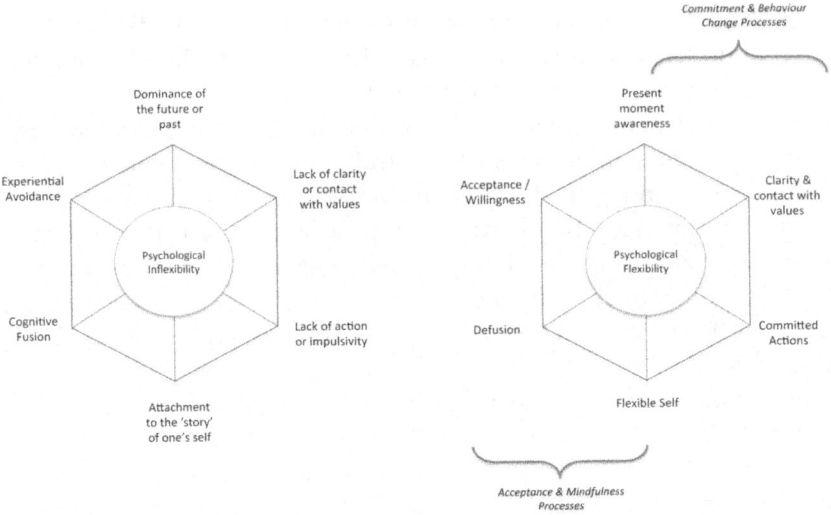

Figure 1: The acceptance and commitment therapy model—processes of inflexibility and psychological flexibility.

(being in touch with what is most deeply cared about in life), committed action (making and keeping commitments to specific values-related actions, big and small), and a sense of self as context (a set of flexible perspective taking skills that involve less attachment to a narrowly defined story about the self). For a book-length explanation of ACT, see Hayes and colleagues (2012).

Psychological flexibility is a concept that appears to be similar to the neuropsychological construct of cognitive flexibility, which is considered to be the "ability to shift avenues of thought and action in order to perceive, process and respond to situations in different ways" (Eslinger & Grattan, 1993, p. 17). These authors describe a number of neural sub-strates for cognitive flexibility and suggest that cognitive inflexibility is a property of neurological impairment, chiefly of the frontal lobes and basal ganglia. Psychological inflexibility, by contrast is considered to be dimensional (i.e., people in the population will vary according to how flexible they are, covering the normal and clinical populations) and contextual (people will behave more flexibly or inflexibly according to specific contexts and patterns of learning history within that context). Given the early stage of development of applying ACT in neurological populations, little is known about how these (and related concepts such as executive function, response inhibition, anticipating consequences, perspective taking, and deficits in empathy) might relate to each other. It is likely that deficits in cognitive flexibility may make it harder for someone to be psychologically flexible. Psychological flexibility requires an individual to attend to sources of behavioural influence upon them, such as thoughts, feelings, and the external environment, to be able to shift perspective from immediate to distal concerns, and to be able to inhibit "automatic" or well-learned responses in favour of responses that lead towards overarching life goals. Given that this skill relies upon holding several competing strands of attention at once, as well as inhibiting behaviour, it is a reasonable hypothesis that neuropsychological problems, such as lack of cognitive flexibility, impairment in executive function, deficits in perspective taking, and deficits in response inhibition will make it harder to implement ACT work in neuropsychological settings.

ACT is however, beginning to be applied with populations with neurological problems. For instance, Sylvester (2011) describes a small study with seventeen adults experiencing the late effects of acquired brain injury (ABI) in childhood. An eight session ACT group produced increased engagement in adaptive behaviours and decreased avoidance, but with no change in threatening cognitions about ABI. Similarly, Bedard and colleagues (2003) describe a mindfulness based stress reduction (MBSR) intervention to improve quality of life following ABI in adulthood. In ten people with mild to moderate ABI, the intervention helped improve quality of life and reduce symptoms of depression. Although MBSR is not the same as ACT, the intervention was described as using MBSR to teach participants "how to approach life with a sense of acceptance, allowing them to move beyond limiting beliefs" (Bedard et al., 2003, p. 724). The small amount of evidence in this field, along with suggestions for the adaptation of ACT to ABI populations is reviewed by both Kangas and MacDonald (2011) and also by Soo and colleagues (2011). Both reviews suggest that ACT could be usefully applied to people with neurological conditions and cognitive and sensory impairment. Finally, given the interest in this field, it is of note that there is a randomised controlled trial of ACT for traumatic brain injury being carried out through the University of Wollongong, Australia (Whiting et al., 2012).

The potential application of ACT to living with MS is a logical step and is supported by both cross sectional and longitudinal research. Ferenbach (2011) for example conducted a cross sectional study of 132 people with MS, investigating relationships between symptoms, distress, and life satisfaction. Results showed that the relationship between symptoms and distress was more strongly influenced by cognitive fusion (i.e., how we stand towards our own thoughts) than by the content of beliefs (e.g., appraisals of illness as uncontrollable and the self as helpless). Similarly, Pakenham and Fleming (2011) showed that after controlling for initial adjustment, demographic factors, and illness variables, higher MS acceptance was predictive of positive adjustment twelve months later, in a sample of 128 people with MS. Finally, Sheppard et al. (2010) describe a pilot study involving fifteen people with MS who took part in a half-day workshop based on ACT principles. Results suggested improvements in depression and quality of life over a period of three months.

In summary, there is evidence that traditional forms of CBT can help people to adapt well to having MS and can be useful in living with the symptoms and associated psychological features. In addition there is evidence that constructs from ACT such as acceptance, cognitive fusion, and contact with values may also be useful targets for treatment, particularly when appraisals of MS are realistic, as may be the case in people with SPMS.

The case presented below describes the application of an ACT approach with a woman with secondary progressive MS. In addition to issues related to MS, two other important features of the work should be introduced. The first is that part of the intervention was a couples-based intervention and the second that an important element of the therapeutic work also focused on the patient's history of childhood maltreatment, and the psychological sequelae of this. ACT has been used with couples, both in general settings (Honarparvaran et al., 2010) and in neurological settings (Williams et al., 2014). In working with couples, ACT is typically used to help couples to respond more flexibly, with greater kindness and love towards their partner, and to focus on shared values, even in the presence of judgements, emotions, disappointments, and so on (see Harris, 2009).

ACT has also been successfully used to treat issues related to childhood trauma, though at present the evidence tends to consist of case studies (e.g., Batten & Hayes, 2005), as well as books and chapters describing extensive clinical experience with this population (e.g., Batten et al., 2005; Walser & Westrup, 2007). As yet there have been no clinical trials of ACT specifically for psychological problems related to traumatic childhood events. In contrast, there have been a number of cross sectional and observational studies that have shown the relevance of psychological flexibility as a treatment target in such populations. For instance, Kingston and colleagues (2010) present a structural equation model (SEM) in a clinical sample of 290 people with a wide range of behaviour problems (e.g., alcohol and substance misuse, aggressive behaviour, binge eating, deliberate self-harm, sexual promiscuity). The SEM showed that the relationship between exposure to trauma in childhood and behaviour problems in adulthood is fully mediated by experiential avoidance, consistent with ACT theory (Kingston et al., 2010). This suggests that although the evidence for using ACT to treat trauma-related problems is not yet well developed, the core process of the ACT model is well supported as a viable treatment target, and that case study evidence suggests it as a promising intervention.

Case study

Introduction

All names and identifying details have been altered to protect the patient's confidentiality. The patient gave consent for the material to be written up for publication. Jane is a sixty-two-year-old lady with secondary progressive MS. She was referred to the treating neuropsychologist (SG) by an occupational therapist to address problems with low mood and anxiety. The neuropsychologist was a member of a multidisciplinary team that specialised in working with people with progressive neurological conditions. In addition to the psychological intervention she was receiving, Jane also had planned two-week admissions to the rehabilitation unit every ten weeks during which she received assessments and support from a number of other professionals, including nursing, occupational therapy, physiotherapy, and medicine. Jane's psychotherapy treatment was begun while an in-patient and continued as an out-patient.

At the point of treatment she was wheelchair dependent, had no functional use of her upper limbs, and mild cognitive difficulties were noted. There was evidence of cognitive slowing and mild difficulties with retaining and recalling verbally presented information. Mental fatigue was noticeable during afternoon sessions and appointments were organised to accommodate for this.

Jane had been married to her husband, Howard, for forty years; they had two adult children and two grandchildren. She lived at home with her husband and was supported by a care package involving two visits a day.

Presenting complaints

Jane was independent and organised prior to the progression of her MS and found it difficult to cope with the severity of her current symptoms and their impact on her sense of control. She was withdrawing from others and finding it difficult to communicate with her husband and adult children. She felt guilty about requiring help from others and of not being able to support her husband with his own chronic health difficulties. She was highly anxious, which exacerbated tension and pain in her shoulders.

History

Jane described a chaotic and abusive childhood involving sexual abuse from her mother's partner and prolonged neglect of her physical and emotional needs. In her adolescence she went to live with her grandmother who was controlling, critical, and uncaring. She moved between a number of homes throughout her childhood that she described as dirty, unsafe, and full of people she did not know. At her grandmother's house she shared a room with an aunt and had little privacy. Jane experienced rejection from her peers, grew up with a sense of being different, and feared that people would judge her if they knew the truth about her life. She was led to believe that she owed other people and ought to please them. She described feeling scared, ashamed, and vulnerable and that she needed to be independent from a young age.

Jane got married when she was twenty-two, raised two children and ran the house as her husband worked away from home. She first experienced MS symptoms when she was thirty-one years old but was not diagnosed until she was forty-nine. At this time she began falling and dragging her left leg, which was noticeable to others and embarrassing for her. She found it difficult to use a wheelchair or accept a catheter due to dignity and embarrassment issues.

Assessment

The first session involved a clinical interview with Jane as the referral specifically related to her low mood and anxiety, and no mention was made of her relationship difficulties. As she highlighted problems with withdrawing from others and her concerns about supporting her husband with his health difficulties, she agreed that a joint assessment session with him could be useful. He had one session alone followed by a joint session in which a treatment plan was developed. The assessment was conducted from an ACT perspective with a focus on the short- and long-term consequences of Jane and Howard's responses to SPMS.

At the beginning of the intervention, Jane's main problems were feeling frustrated by her MS, disempowered, and missing her role as a wife, mother, and grandmother. She had always had problems relaxing but had been less troubled by her anxiety and low mood when she had been physically able and in charge of running the house. She now felt like a burden on others and that there had been a role reversal between herself and her daughter. She currently felt unable to cope and that her low mood and distress was having a significant impact on her life.

Howard felt overwhelmed by caring and the need to stay on top of things at home. He had hearing difficulties and chronic bowel problems that caused pain, impacted on functioning, and required regular hospital appointments. He was also involved in caring for their granddaughter and took her to and from nursery three days a week. Both identified poor communication as a problem.

Measures

The Acceptance and Action Questionnaire (AAQ-II) (Bond et al., 2011) is a single-domain measure of psychological inflexibility. It consists of seven statements such as "I worry about not being able to control my worries and feelings" and respondents indicate the extent to which these statements apply on a scale from one (never true) to seven (always true). Scores range from seven to forty-nine, with higher scores indicating greater inflexibility. The AAQ-II has been studied with over 2000 participants and has demonstrated acceptable psychometric properties (Bond et al., 2011). Acceptance has also been associated with better adjustment to MS (Pakenham & Fleming, 2011) and was therefore deemed to be an important treatment variable to measure.

The Hospital Anxiety and Depression Scale (HADS) (Zigmond & Snaith, 1983) is a fourteen item measure of anxiety and depression. Items include "I feel tense and wound up" and "I still enjoy the things I used to enjoy". It aims to avoid measuring aspects of the condition that are common somatic symptoms of illness such as fatigue and insomnia and as such is appropriate for use in the MS population. Each item is scored from zero to three and scores range from zero to twenty-one on each subscale.

Initial assessment results

At the beginning of the intervention, Jane's scores on the HADS and AAQ-II (see Table 1 on p. 70) indicated that she was experiencing moderate levels of anxiety and mild levels of depression. Her AAQ-II score was thirty-four out of forty-nine, highlighting moderate levels of avoidance/inflexibility. This was consistent with the personal reports given by Jane during clinical interview.

Case conceptualisation

Initially Jane's low mood and anxiety were conceptualised as being the consequence of avoidance strategies, aimed at reducing her exposure to distressing thoughts about her MS such as "I shouldn't be a burden on other people". Based on the six processes of psychological flexibility (Hayes et al., 2004) it was hypothesised that in order to reduce the distress that arose from requiring care and not being actively involved in housework, Jane had stopped asking her husband for the things that she needed and would spend time alone in her bedroom to avoid seeing clutter in the house that she wished was tidied away. As a consequence she felt that people did not understand what she was going through; that she did not have her needs met, and that she was lonely. This low mood and isolation in turn limited her engagement in valued activities.

After six sessions, Jane disclosed information about her childhood that was hypothesised to be influencing the way she coped with her MS. Reformulation of the case proposed that Jane was influenced by thoughts and beliefs that had been present from her early years including "I am different and vulnerable", "other people will judge me", and "people will let me down" (cognitive fusion). Jane's sense of self was focused on shame and inadequacy and led to fears of being judged and let down by others (attachment to a narrow "self story"). Another main theme was around control. She was unable to control many aspects of her childhood, making her vulnerable to anxiety and distress. In early adulthood she was able to be in control of raising her children and keeping her home, leading to a reduction in anxiety. She now experienced a lack of control over her MS, its symptoms, and her need for care. The similarities between her childhood and her life with MS caused her to feel vulnerable and frightened. She struggled to see herself as being separate from these thoughts (self as context) and as a result felt very threatened.

There was also evidence of experiential avoidance from a young age. The abuse she experienced had caused a number of traumatic memories that she dealt with as an adolescent by trying to avoid disapproval and further harm by becoming independent and not relying on other people. In adulthood, before her MS symptoms progressed, she avoided these mental experiences by keeping her home very tidy, having regular routines with her children, and managing independently. As her MS progressed, however, she was no longer able to maintain her home to her high levels of cleanliness and became dependent on other people for support. This increased her sense of vulnerability, exacerbated her anxiety about being judged, and caused discomfort about clutter in the house. In response she again employed an avoidance strategy by withdrawing to her own bedroom and communicated less with her family. This avoidance strategy was reinforced by reducing her exposure to triggers of painful mental events, leading to short-term reductions in distress. In addition, by not talking to her relatives she was able to maintain a sense of not depending on others and reduced her fear of being judged by not disclosing information about herself. From a resilience perspective it was acknowledged that being independent during her adolescence served an adaptive purpose and helped to keep her safe, however, she was continuing to employ the same strategy, even though the current dangers were greatly reduced.

Course of treatment

Based on the case conceptualisation above, it was felt that Jane could benefit from an ACT intervention that focused on (a) a workability analysis of the strategies she was currently using, (b) a reduction of experiential avoidance, and (c) the pursuit of valued activities.

The intervention involved three distinct phases that totalled ten sessions over a six month period. The first part consisted of three assessment sessions: a clinical interview with Jane, one with Howard, and a joint session together. The second phase involved three joint sessions with the couple and focused on the thoughts and feelings the couple had about MS and how their avoidance of those mental events were obstacles to them pursuing their values. The final phase consisted of four sessions with Jane alone and focused on altering the relationship between childhood experiences and her current responses to MS.

A number of techniques were introduced, beginning with a work-ability analysis that explored the strategies the couple used to manage their anxiety and how successful they were. Howard acknowledged that he became focused on practical tasks and avoided emotional issues. In the short-term this allowed him to feel that he was keeping on top of household chores, however in the longer-term it caused fric-tion between the couple, left Jane feeling excluded from activities she previously managed, and prevented them from talking about emotional issues. The couple acknowledged that these strategies had not been effective.

Jane was able to identify that she withdrew from others, which in the short-term allowed her to feel less of a burden and to avoid emotional conversations. In the longer-term, however, she felt discon-nected and alone.

In line with Harris (2012), the couple were helped to notice, name, and step back from difficult thoughts and feelings as opposed to trying to control or avoid them. Defusion exercises including leaves on the stream (Hayes et al., 2012, pp. 255–258) were used to encour-age the couple to practise seeing the passing nature of mental events and to help them feel less enmeshed with them. They were encouraged to name some of the thoughts for example, "there's the 'must keep busy' thought again". The couple realised that their anxieties about the past and the future prevented them from noticing what was happening in their immediate environment. Mindfulness exercises, including focusing on their breath, were introduced during sessions and practise at home was encouraged, using audio instructions recorded on to CD. Audio recordings of these exercises are available for free download at: http://contextualscience.org/david_gillanders_training_page.

To support the couple in identifying and engaging in committed action, time was spent discussing their individual and shared values. Both of them said they would like to communicate more openly with one another and Jane identified that she would like to be more connected to her children. Due to her mild cognitive deficits, specific goals were identified including sitting together after breakfast and trying to make joint decisions about household affairs. Jane also committed to speaking to her children about her MS and to answer their questions openly. We identified potential barriers to pursuing these values and a written summary was produced for the couple to refer to at home.

At the end of this phase of the intervention, the couple felt that they were more able to notice their thoughts and step back from them. They were pursuing their values of communicating better with one another and Jane felt she was more open with her children.

The third phase of the intervention was focused on Jane alone and we spent time talking about her childhood and the thoughts and feelings that continued to upset her. Techniques from earlier in the intervention were employed including noticing, naming, defusing, mindfulness, and choosing valued activity, even in the presence of thoughts, feelings, beliefs, and memories related to abusive experiences. These strategies are consistent with the case studies and clinical manuals that describe the successful application of ACT for problems related to traumatic childhood experiences (Batten & Hayes, 2005; Batten et al., 2005; Walser & Westrup, 2007).

Metaphors are commonly used in ACT and one that proved helpful for Jane was the "passengers on the bus" (Hayes et al., 2012, p. 250). In this metaphor, the client is the driver of the bus and is trying to steer it in the direction of their values. In Jane's case, as the bus moved towards a value such as being more open with her children, passengers would get up from their seats, move to the front, and say things like "you'll only overwhelm them if you tell them" and "they'll judge you if they know the truth", which caused her to want to steer back towards the position of remaining private and guarded. This metaphor helped Jane to appreciate that by getting entangled in her negative thoughts, it was difficult to pursue her values. By observing these "passengers" and personifying them, for example naming and creating an image of "The Judge" who represents her fears of people judging her, Jane was more able to defuse from their intensity and instead chose how she wanted to respond, for example being more open with her children despite to presence of "passengers" who suggested she ought not to.

Adaptations to the intervention

Jane's mild cognitive difficulties meant that a number of adaptations were made to delivering ACT. As described above, the core treatment target of the ACT model is psychological flexibility. This concept refers to the ability to notice mental events in the moment and to flexibly choose how to respond. This conscious awareness and choice differs from the default use of previously employed strategies that were hypothesised to function to avoid distress. The neuropsychological concept of cognitive flexibility is considered to be the "ability to shift avenues of thought and action in order to perceive, process and respond to situations in different ways" (Eslinger & Grattan, 1993). As described above, there is reason to believe that deficits in cognitive flexibility would impact upon a client's psychological flexibility. Jane demonstrated a good ability to think flexibly about her past and to make values based choices in therapy. To maximise her skills in this area, a structured approach was used in which the topic to be considered was clearly stated and the pros and cons of different options were written down to reduce the impact on her working memory.

Jane also benefited from the chunking of information and repetition. A diagram was collaboratively developed to visually portray the links between her mental events, avoidance strategies, and their impact on her behaviour and well-being. Written summaries were provided after sessions to promote retention and specific goals were set during sessions to give clarity and structure to homework exercises.

Finally, the defusion function of the passengers on the bus metaphor was made more concrete by drawing an individualised cartoon depicting monsters on a bus. Similar cartoons are available to download at: http://contextualscience.org/visual_aids.

Clinical outcome

By the end of therapy, Jane had a better understanding of her difficulties and even though this was not a treatment target, she felt calmer and less tense. She used to go "deeper and deeper" into her memories about her childhood but now felt able to step back from them and let them pass. She was also committed to living differently with her history, and had allowed herself to feel angry about the impact her childhood had inflicted on her as an adult.

Jane had managed to go out for a meal for her granddaughter's birthday and allowed her daughter to feed her in public. She was more accepting of mess in the house, seeing it as being the result of having family around her. She felt like a good mum and recognised that people think she listens and cares. She felt closer to her four-year-old granddaughter who was now choosing to spend time with Jane when she was at their house. She said she felt embedded within the family. These behavioural changes show that she was more successfully pursuing her values, even with SPMS and its challenges.

Table 1 and Figures 2 and 3 show Jane's HADS and AAQ-II scores throughout the intervention. Her scores on the HADS shows that her anxiety and depression scores had dropped to the normal range by the end of therapy. Her AAQ-II scores (Figure 3) increased at time point two and on discussion, this was found to be because of ongoing struggles with traumatic memories from childhood. At the end of the intervention, this had decreased considerably, suggesting a greater acceptance of her mental events and a greater ability to pursue valued activities. Her final score of twenty-five is below the threshold of twenty-eight reported by Bond and colleagues (2011) as indicative of psychological distress on the Beck Depression Inventory (BDI-II) (Beck et al., 1996) and the Global Severity Index of the Symptom Checklist—Revised (SCL-90-R; Derogatis, 1994). In addition, Reliable Change Index (RCI) and Clinically Significant Change Index (CSC) scores were calculated for the pre to follow up changes on the HADS scales and the AAQ-II. The method used for calculation is based upon

Table 1: Measures over the intervention

	Pre	*Mid*	*Final*	*Change*	*RCO*	*CSC*	*Interpretation*
AAQ-II	34/49*	39/49	25/49	9	1.4	≤ 19	Improved
HADS-A	13	9	7	6	2.4	≤ 7.5	Recovered
	moderate	mild	normal				
HADS-D	9	9	4	5	1.8	≤ 5	Recovered
	mild	mild	normal				

Note: RCI = Reliable Change Index, CSC = Clinically Significant Change. *Higher scores indicate inflexibility/avoidance, lower scores indicate greater psychological flexibility, descriptors for the HADS are based on Jones and colleagues (2012). Interpretation of clinically significant change is based upon Jacobson and Truax (1991).

Jacobson and Truax (1991), but used the Leeds Reliable Change Index Calculator (LRCIC) (Agostiniset al., 2008).

The LRCIC is an automated excel file, into which users can insert the pre- and post-treatment scores of an individual, along with known normative data for the scale (Cronbach's α, mean, and SD for clinical and normative samples). The calculator will then provide calculations for the RCI and CSC. For the current case, normative data for the HADS was derived from Crawford et al. (2001). HADS reference data for a large MS sample was derived from Jones and colleagues (2012).

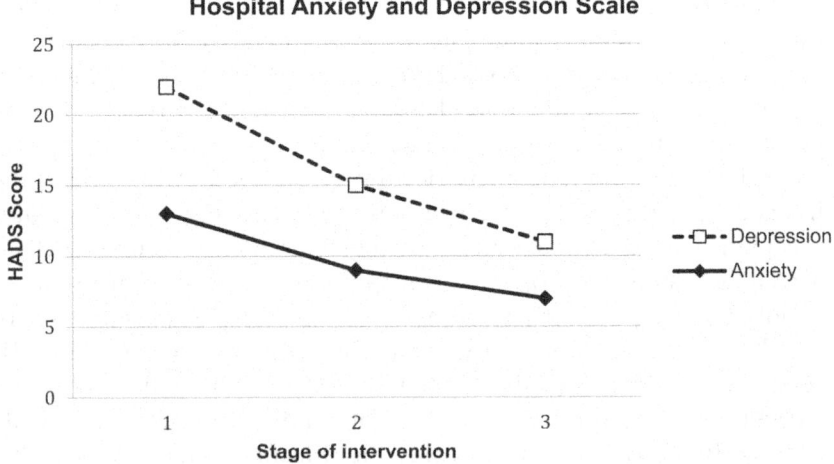

Figure 2: Anxiety and depression scores on the HADS at the beginning, middle and end of the intervention.

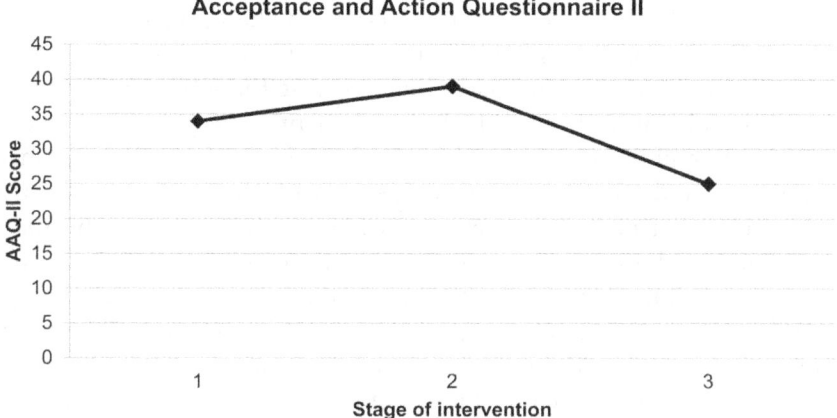

Figure 3: Psychological inflexibility scores on the AAQ-II at the beginning, middle and end of the intervention.

Normative data for the AAQ-II was derived from Bond and colleagues (2011) and norms for a mixed MS sample from Ferenbach (2011). For both scales, Jacobson and Truax's criterion c for CSC was used: that the final score should fall in a region where it is more likely to belong to the non-treatment seeking population than the treatment seeking population (Jacobson & Truax, 1991). For all measures, pre- to post-treatment change scores exceeded the RCI value, indicating that changes were reliable and unlikely to be due to measurement error. In addition, clinically significant changes were shown on both the HADS scales, indicating that Jane had recovered. By the end of treatment, her score on the AAQ-II was not yet below the cut off for clinically significant change, and so Jane could be considered improved but not recovered, using this more stringent method for evaluating clinically significant change.

Treatment implications of the case

To date there have been few case studies examining the clinical utility of ACT for MS. This case has shown that it can be usefully applied with an individual with SPMS who has significant physical disability. By adapting the approach through the use of repetition, written summaries, more specific recommendations than is typical in ACT, and diagrams to facilitate therapy, it could be usefully applied to individuals with mild cognitive difficulties. It was also able to incorporate work with a couple where relationship strain and shared values were evident. The approach was also able to address themes commonly seen in the MS population including challenges to self-identity, vulnerability, and control. These were particularly pertinent issues given Jane's experiences in childhood and facilitated increased awareness, acceptance, resilience, and self-compassion at this stage of her life. There were also changes in her perceptions about her MS. She initially thought that if she had tried harder in the early stages of the condition, that she would not have lost the function of her hands. During the course of therapy she became more able to recognise that she was not responsible for the progression of her MS and experienced a change in self-schema that allowed her to hold the belief that "I am able live with this condition". What is particularly noteworthy is that these beliefs changed despite not being a focus of intervention. We suggest that rather than seeing belief change as the mechanism of action (as would be the case in traditional cognitive therapy), these new appraisals are secondary to actually living more flexibly with the condition.

Limitations

Despite Howard being involved in the intervention, only verbal self-report of his well-being was elicited. In hindsight it would have been useful to measure this using standardised instruments. A second limitation was the omission of a condition specific measure of acceptance. Pakenham and Fleming (2011) report on a sixteen item measure of MS acceptance (MSAQ). Initially it was felt that the AAQ-II alone was sufficient as it was a more well-established measure than the MSAQ. On discussion with Jane, however, she reported feeling more accepting of MS but not her early childhood memories and the use of both the MSAQ and the AAQ-II may have been able to measure this change. Future ACT interventions with this population should consider the use of condition specific measures.

Recommendations to clinicians and students

Further research is needed into the effects of childhood trauma on well-being and adjustment in people with MS. This includes exploring the role of resilience as a helpful strategy, as well as acknowledging and addressing the impact of abuse on psychological well-being when living with this progressive condition. The results of this case study suggest that there could be utility in developing a structured ACT based protocol for individuals and couples living with MS, with specific treatment targets based on stage of disease progression. Testing the effectiveness of such protocols in larger group based studies could help to increase choice of psychological interventions available to people with MS.

Interestingly, while it is reasonable to hypothesise that problems with cognitive flexibility, response inhibition, and deficits in perspective taking may make it harder for someone to become more psychologically flexible, it is also reasonable to test the hypothesis that ACT interventions designed to help people to slow down, increase awareness, take perspective, and change habitual modes of responding, may modify the behavioural effects of such neuropsychological problems.

Finally, monitoring of changes in belief content and cognitive fusion in response to interventions designed to target cognitive change (e.g., cognitive therapy) or cognitive fusion (e.g., ACT) could help to clarify the mechanisms of action of cognitive and behavioural therapies. In addition, such future research could clarify the relationship between content and fusion. Such an approach will not only likely lead to more efficient and targeted psychological treatment, it will also shape our theoretical understanding of cognition, affect, and behaviour.

References

Agostinis, A., Morley, S. J., & Dowzer, C. N. (2008). The Leeds Reliable Change Index Calculator. Available at: http://medhealth.leeds.ac.uk/info/618/clinical_psychology_dclinpsychol/797/leeds_reliable_change_index. Last accessed 24 April 2014.

Alford, B. A., & Beck, A. T. (1997). *The Integrative Power of Cognitive Therapy*. New York: Guilford.

Batten, S. V., & Hayes, S. C. (2005). Acceptance and commitment therapy in the treatment of comorbid substance abuse and post-traumatic stress disorder: a case study. *Clinical Case Studies*, 4(3): 246–262.

Batten, S. V., Orsillo, S. M., & Walser, R. D. (2005). Acceptance and mindfulness-based approaches to the treatment of posttraumatic stress disorder. In: Susan M. Orsillo & Lizbeth Roemer (Eds), *Acceptance and*

Mindfulness-based Approaches to Anxiety: Conceptualisation and Treatment. New York: Springer.

Beck, A. T., Steer, R. A., & Brown, G. K. (1996). *Manual for the Beck Depression Inventory-II*. San Antonio, TX: Psychological Corporation.

Beck, J. S. (2011). *Cognitive Behavior Therapy: Basics and Beyond* (2nd edn). New York: Guilford.

Bedard, M., Felteau, M., Mazmanian, D., Fedyk, K., Klein, R., Richardson, J., Parkinson, W., & Minthorn-Biggs, M. (2003). Pilot evaluation of a mindfulness-based intervention to improve quality of life among individuals who sustained traumatic brain injuries. *Disability & Rehabilitation, 25*(13): 722–731.

Blackledge, J. T. (2007). Disrupting verbal processes: cognitive defusion in acceptance and commitment therapy and other mindfulness-based psychotherapies. *The Psychological Record, 57*: 555–576.

Bonanno, G. A. (2004). Loss, trauma, and human resilience: have we underestimated the human capacity to thrive after extremely aversive events? *American Psychologist, 59*: 20–28. doi: 10.1037/0003-066X.59.1.20

Bond, F. W., Hayes, S. C., Baer, R. A., Carpenter, K. C., Guenole, N., Orcutt, H. K., Waltz, T., & Zettle, R. D. (2011). Preliminary psychometric properties of the Acceptance and Action Questionnaire-II: a revised measure of psychological flexibility and acceptance. *Behavior Therapy, 42*: 676–688.

Coker, A. L., Follingstad, D., Garcia, L. S., Williams, C. M., Crawford, T. N., & Bush, H. M. (2012). Association of intimate partner violence and childhood sexual abuse with cancer-related well-being in women. *Journal of Women's Health, 21*(11): 1180–1188.

Crawford, J., Crombie, C., & Henry, J. (2001). Normative data for the HADS from a large non-clinical sample. *British Journal of Clinical Psychology, 40*: 429–434.

Derogatis, L. R. (1994). *SCL-90-R: administration, scoring and procedures manual*. Minneapolis, MN: National Computer Systems.

Eslinger, P. J., & Grattan, L. M. (1993). Frontal lobe and frontal-striatal substrates for different forms of human cognitive flexibility. *Neuropsychologia, 31*(1): 17–28.

Ferenbach, C. (2011). Process of psychological adjustment to multiple sclerosis: comparing the roles of appraisals, acceptance, and cognitive fusion. Unpublished Doctoral Thesis, University of Edinburgh. Available at: www.era.lib.ed.ac.uk/handle/1842/6289

Harris, R. (2009). *ACT with Love*. Oakland, CA: New Harbinger.

Harris, R. (2012). *The Reality Slap*. London: Constable & Robinson.

Hayes, S. C., Strosahl, K. D., & Wilson, K. G. (2012). *Acceptance and Commitment Therapy: The Process and Practice of Mindful Change* (2nd edn). New York: Guilford Press.

Hayes, S. C., Strosahl, K. D., Bunting, K., Twohig, M., & Wilson, K. (2004). What is acceptance and commitment therapy? In: S. Hayes & K. Strosahl (Eds.), *A Practical Guide to Acceptance and Commitment Therapy* (pp. 3–29). New York: Springer.

Honarparvaran, N., Tabrizy, M., Navabinejad, Sh., Shafiabady, A., & Moradi, M. (2010). The efficacy of acceptance and commitment therapy (ACT) training with regard to reducing sexual dissatisfaction among couples. *European Journal of Social Sciences, 15*: 166–172.

Hopwood, M., & Treloar, C. (2008). Resilient coping: applying adaptive responses to prior adversity during treatment for hepatitis C infection. *Journal of Health Psychology, 13*: 17–27. doi:10.1177/1359105307084308

Jacobson, N. S., & Truax, P. (1991). Clinical significance: a statistical approach to defining meaningful change in psychotherapy research. *Journal of Consulting and Clinical Psychology, 59*(1): 12–19.

Jones, K. H., Ford, D. V., Jones, P. A., John, A., Middleton, R. M., Lockhart-Jones, H., Osborne, L. A., & Noble, J. G. (2012). A large-scale study of anxiety and depression in people with multiple sclerosis: a survey via the web portal of the UK MS Register. PLoS One. 7(7): e41910. doi: 10.1371/journal.pone.0041910

Kangas, M., & MacDonald, S. (2011). Is it time to act? The potential of acceptance and commitment therapy for psychological problems following acquired brain injury. *Neuropsychological Rehabilitation: An International Journal, 21*(2): 250–276.

Kingston, J., Clarke, S., & Remington, B. (2010). Experiential avoidance and problem behavior: a mediational analysis. *Behavior Modification, 34*(2): 145–63. doi:10.1177/0145445510362575

Moss-Morris, R., Dennison, L., & Chalder, T. (2010). *Supportive Adjustment for Multiple Sclerosis (saMS): an eight-week CBT programme manual.* London: MS Society. (retrieved 28th Feb 2014 from www.mssociety. org.uk/sites/default/files/Documents/Professionals/SAMS%20Manual %20GI22%201210%20-%20web.pdf).

Moss-Morris, R., Dennison, L., Landau, S., Yardley, L., Silber, E., & Chalder, T. (2013). A randomized controlled trial of cognitive behavioral therapy (CBT) for adjusting to multiple sclerosis (the saMS trial): does CBT work and for whom does it work? *Journal of Consulting and Clinical Psychology, 81*: 251–262. doi:10.1037/a0029132

Mutch, K. (2010). In sickness and in health: experience of caring for a spouse with MS. British *Journal of Nursing, 19*(4): 214–220.

Pakenham, K. I., & Fleming, M. (2011). Relations between acceptance of multiple sclerosis and positive and negative adjustments. *Psychology & Health, 26*(10): 1292–1309. doi:10.1080/08870446.2010.517838

Roy, A., Roy, M., & Goldman, D. (2011). Childhood trauma and depressive symptoms in type 1 diabetes. *Journal of Clinical Psychiatry, 72*: 1049–1053.

Sheppard, S. C., Forsyth, J. P., Hickling, E. J., & Bianchi, J. (2010). A novel application of acceptance and commitment therapy for psychosocial problems associated with multiple sclerosis. *International Journal of MS Care, 12*: 200–206. doi:10.7224/1537-2073-12.4.200

Soo, C., Tate, R. L., & Lane-Brown, A. (2011). A systematic review of acceptance and commitment therapy (ACT) for managing anxiety: applicability for people with acquired brain injury? *Brain Impairment, 12*(1): 54–70. doi:10.1375/ brim.12.1.54

Sylvester, M. (2011). Acceptance and commitment therapy for improving adaptive functioning in persons with a history of pediatric acquired brain injury. University of Nevada, Reno, ProQuest, UMI Dissertations Publishing.

Vukusic, S., & Confavreux, C. (2003). Prognostic factors for progression of disability in the secondary progressive phase of multiple sclerosis. *Journal of Neurological Science, 206*: 135–137.

Walser, R. D., & Westrup, D. (2007). *Acceptance & Commitment Therapy for the Treatment of Post-Traumatic Stress Disorder & Trauma-Related Problems: A Practitioner's Guide to Using Mindfulness & Acceptance Strategies*. Oakland, CA: New Harbinger.

Whiting, D. L., Simpson, G. K., Mcleod, H. J., Deane, F. P., & Ciarrochi, J. (2012). Acceptance and commitment therapy (ACT) for psychological adjustment after traumatic brain injury: reporting the protocol for a randomised controlled trial. *Brain Impairment, 13*(3): 360–376.

Williams, J., Vaughan, F., Huws, J., & Hastings, R. (2014). Brain injury spousal caregivers' experiences of an acceptance and commitment therapy (ACT) group. *Social Care and Neurodisability, 5*(1): 29–40. doi:10.1108/SCN-02-2013-0005

Zigmond, A. S., & Snaith, R. P. (1983). The hospital anxiety and depression scale. *Acta Psychiatrica Scandinavica, 67*: 361–370.

CHAPTER FOUR

Soothing the injured brain with a compassionate mind: building the case for compassion focused therapy following acquired brain injury*

Fiona Ashworth

Introduction

T his chapter is divided into four key sections in order to provide a coherent overview for the rationale of using compassion focused therapy (CFT) to ameliorate psychological distress following acquired brain injury (ABI); the first section focuses on shame and self-criticism in association with psychological distress and treatment; the second section provides an introduction to CFT including the key components of the approach and the evidence base. The third section provides special considerations for CFT following ABI, and the final section provides a case example of applying CFT following ABI with closing comments.

Psychological distress and its treatment following ABI

Psychological difficulties (hereafter referred to as psychological distress) such as depression, anxiety, and maladjustment are significant problems following ABI (e.g., Robinson & Spalletta, 2010; Seel & Kreutzer, 2003; Soo & Tate, 2007). Co-morbidity of psychological disorders is also problematic, with estimates at forty-four per cent and

---

* Originally published in 2014 in *Neuro-Disability & Psychotherapy*, 2(1/2): 41–79.

above (Hibbard et al., 1998; Soo & Tate, 2007). Despite this there is mixed evidence to support the efficacy of psychological interventions in this population. The focus on treatment post-ABI has often been disorder specific, and most efficacy studies conducted have used cognitive behavioural therapy (CBT). Some studies show good results (e.g., Bradbury et al., 2008; Hsieh et al., 2012), however, others indicate mixed results (e.g., Hodgson et al., 2005; Rasquin et al., 2009) and negative outcomes have also been reported (e.g., King, 2002). A recent review of the evidence base for CBT for anxiety and depression following ABI (Waldron et al., 2013) conclude that there is some support for the effectiveness of CBT, but it is not a panacea. Block and West (2013) reviewed the evidence for psychotherapeutic interventions following traumatic brain injury (TBI), again concluding some support for CBT as well as narrative and holistic approaches.

More recently evidence is emerging for the value of mindfulness-based interventions in the treatment of psychological distress (and in some cases cognitive impairments) following ABI. There are relatively few studies in this area but the results to date have been promising (e.g., Bedard, et al., 2013). However, a number of recent studies suggest that it is specific attitudinal factors including having a compassionate and non-judgmental stance towards one's thoughts, feelings, and behaviours that is more influential in promoting change towards positive well-being and mental health, rather than the focused attention and observation of one's internal experiences that is the primary focus of mindfulness based interventions (e.g., Miron, 2013; Van Dam et al., 2011). This will be discussed in the section on compassionate mind training (CMT) in this chapter.

Although the evidence base for the treatment of psychological distress following ABI is growing it is still limited by a number of factors including sample size, the relatively small number of studies, selection criteria, and varied treatment protocols. It is further complicated by the heterogeneity of psychological distress following ABI and the cognitive impairments arising as a result of the ABI. Given that CBT requires considerable cognitive ability and effort, it can be criticised for expecting some clients to engage with tasks that require the very abilities that may be impaired.

Shame and self-criticism as transdiagnostic processes

In the last decade in mainstream (non-brain injured) mental health research, there has been a drive towards understanding key under-lying processes (i.e., transdiagnostic processes) that may be contribut-ing to and maintaining psychological disorders (e.g., Gilbert & Procter, 2006; Harvey et al., 2004). A number of transdiagnostic processes have been proposed including cognitive, emotional, and behavioural. This has led to the emergence of research to investigate whether targeting such processes for intervention may lead to positive outcomes across disorders (e.g., Gilbert & Procter, 2006; McEvoy et al., 2009). The issue of heterogeneity and co-morbidity of psychological disorders follow-ing ABI lends support to a transdiagnostic approach as emerging evidence implies shared underlying processes across disorders and amelioration of co-morbid symptomatology following transdiagnostic interventions (e.g., Dozois et al., 2009; Gilbert & Procter, 2006; Judge et al., 2012).

Shame and self-criticism are emerging as pervasive (transdiagnos-tic) features contributing to and maintaining a range of different psychological disorders (e.g., Gilbert & Irons, 2004, 2005; Tangney & Dearing, 2002; Zuroff et al., 2004). CFT (Gilbert, 2006, 2009, 2014) has been developed as a transdiagnostic intervention aimed specifically at targeting shame and self-criticism, and emerging evidence shows that CFT is effective in the treatment of people with high levels of shame and self-criticism (e.g., Braehler et al., 2013; Gilbert & Procter, 2006; Lucre & Corten, 2012).

Shame is an experience that is not commonly discussed or described as central to many therapies. Shame is a subjective affective experience, seen as a response to how the self is negatively perceived/evaluated in the real (or imagined) mind of the self or the other (Goss et al., 1994). Shame plays a central role in CFT (Gilbert, 2002); it is linked to the fact that humans want to create positive feel-ings about the self in the mind of others (Gilbert, 2007). If we are sensi-tive to negative feeling and thoughts about ourselves in the minds of others, we can be vulnerable to external shame (Gilbert, 1997, 1998). Thus if we experience the self in the mind of the other as rejected, unworthy, or vulnerable, this can make the social world unsafe. With internal shame the focus on attention is inwardly on the negative eval-uation of the self as inadequate, or flawed in our own mind.

Gilbert (2010a) highlights key defensive/protective strategies in response to the experience of shame; the first is an internalised response of taking a submissive strategy linked with self-criticising.

Shame-proneness is significantly associated with self-criticism (Gilbert & Miles, 2000). Self-critics have a constant negative perception of themselves and others, and this process may well predispose them to psychopathology after the experience of a stressful/traumatic life event. Another defensive response to shame is an externalised response, for example, being aggressive, angry, or attacking towards others. These responses are not necessarily conscious, but instead reflect phenotypic variations (Gilbert, 2010a).

Shame and self-criticism can require particular attention in psychotherapy as they can play significant roles for an individual including having a vulnerability to psychological distress, viewing the self negatively, developing harmful coping response, and having problems in relating to others. Self-criticism may be a process that narrows an individual's ability to be open and explore their own feelings as they lack social safeness and may feel ashamed (Gilbert, 2010a). Defensive, shame driven responses can also launch direct attacks on the self, the other (therapist), and the process of therapy itself.

Extensive research indicates that heightened self-criticism is significantly associated with a range of psychological disorders (e.g., Cox et al., 2004; Dunkley et al., 2009; Fennig, et al., 2008; Hutton et al., 2013; Luyten, et al., 2007). Vulnerability to self-criticism can lead to an increased stress response as well as decreased social support (e.g., Priel & Shahar, 2000). Self-criticism has been found to be associated with greater symptom severity in depression (Luyten et al., 2007) and a poorer response to psychotherapy including CBT (e.g., Rector et al., 2000). Self-criticism has been linked with higher levels of distress in women with cancer (Campos et al., 2012).

Shame and self-criticism following ABI

The author's clinical experience indicates that self-criticism and shame can be common experiences associated with distress following ABI, although they may predate the injury or evolve as a result of the consequences of the injury itself (as defensive/protective coping strategies). The trend to do research in the area of self-criticism and shame in non-brain injured mental health populations is relatively recent, and unsurprisingly research of this type post-ABI is sparse. Non-empirical clinical case studies and literature regarding psychological adjustment following ABI make reference to self-criticism and self-blame (Ashworth et al., 2011; Prigatano, 1999; Shields & Owns-worth, 2013). Ownsworth and Oei (1998) conducted a study of depression post-TBI, which showed that self-criticism was significantly associated with depression in this population. Research indicates that self-blame as a coping strategy after TBI is associated with higher levels of anxiety and depression (Curran et al., 2000). This was replicated by Anson and Ponsford (2006) where coping characterised by avoidance, wishful thinking, self-blame, and worry were associated with higher levels of anxiety, depression, and psychosocial dysfunction. Research into men's emotional experiences following TBI indicates that they can experience significant shame within the social context of their injury and defensive strategies in response to this include self-criticism, avoidance, and submissive behaviour (Freeman et al., 2014). Qualitative research by Nilsson and colleagues (1997) indicated that difficulties with everyday life following stroke elicited shame as a result of perceptions of failure. Negative perceptions regarding discrepancies between pre- and post-stroke sense of identity were also found to lead to experiences of shame (Dowswell et al., 2000). A number of others studies provide evidence for the role of shame and self-criticism in psychological distress following ABI (e.g., Hagger, 2011; Jones & Morris, 2013).

Compassion focused therapy (CFT)

CFT was initially developed for individuals with high shame and self-criticism associated with chronic mental health problems (Gilbert, 2006). Individuals who experience them can struggle to feel soothed, reassured, or safe in response to threat (Gilbert & Procter, 2006). Conversely, higher levels of self-compassion are associated with lower levels of psychopathology (MacBeth & Gumley, 2012). Therefore, for those individuals who are high in self-criticism and shame, one key aim of CFT is to help an individual refocus on developing a self-compassionate approach.

CFT has its roots in evolutionary psychology, neuroscience, and social psychology approaches, in relation to the psychology and neurophysiology of caring (Gilbert, 2005, 2009, 2014). In CFT, a useful definition of compassion comes from Buddhism: "Compassion is a deep sensitivity to the suffering of self and others, with a deep commitment to try to relieve it" Dalai Lama (1995). There are two key aspects to this: the first is the motivation, competencies, and abilities to notice, engage, and tolerate and make sense of suffering in the self and others, rather than denying or avoiding it. The second is action focused, that is, this involves having the skills and wisdom to know what to do about this suffering (Germer & Siegel, 2012). Although linked, compassion should not be confused with empathy, which is defined as feeling with someone, this is sharing another person's emotion (Singer & Lamm, 2009); empathy can be seen as one element of developing a compassionate mind. Compassion can be viewed as "flow" in three directions; (1) compassion towards others or another, (2) feeling compassion from others towards the self; and (3) compassion that we direct inwardly, towards the self (self-compassion).

CFT is based on an understanding that we have social mentalities embedded in basic social motivational systems (e.g., to form ranks, seek out sexual partners, live in groups, help others, care for family) and different functional emotional systems (e.g., to respond to threats, seek out resources, and contentment/safeness). This focuses on our contextual and relational processing systems and acknowledges that these motivational systems may be operating outside conscious awareness (Gilbert, 2014). When these systems are activated they organise a range of psychological functions including attention, emotion, cognitions, and behaviour in pursuit of that particular goal or motive, hence referred to as a mentality. They also prepare us for communicative and interactional processes and reciprocal relationships (Gilbert, 2005, 2014). Research indicates that the human brain is exceedingly evolved for social processing, such as responding to care, threat, and competition. Research shows that many psychological problems can be entrenched in social relational difficulties. Specifically, difficulties in feeling cared for by others, in having a caring interest in others, and having a caring relationship towards oneself. Supporting clients with these different issues can address a range of problems linked with shame and self-criticism.

Key components of CFT

Essentially, CFT has five key components, although these are not necessarily linear, which are outlined in Table 1 (Gilbert, 2014) and

Table 1: Key components of compassion focused therapy (Gilbert, 2014)

Key elements of CFT
Psycho-education focusing on de-shaming and depersonalising, which focuses on the evolved brain and how it is a "tricky brain". The recognition that much of what goes on in our minds and lives is "not our fault". The three circles model of affect regulation is also explained in the individual's context.
A formulation process where individuals learn about how their early life experiences created coping strategies to defend against threat, as well as drive-excitement based and affiliative soothing strategies and capacities. These are explained as being both internally (how we interact within ourselves and regulate our own emotions) and externally directed (how we interact with others based on what we believe is going on in their minds). This formulation process focuses on understanding historical context, key fears that may have arisen, protective strategies that may have been used to defend against these fears as well as their unintended consequences. Much time may be spent working on trauma memories, which are central to the sense of self and experiences.
Cultivating and building compassionate capacities, by working with affiliative emotions, and learning to practice parasympathetic activation: for example, through breathing and imagery.
Building compassionate capacity around the sense of identity (compassionate self) with behavioural practices. How to take a compassionate perspective and explore what is helpful; what will be the focus of the practice, what will the person cultivate within themselves during the therapy journey?
As these practices develop, the person can be helped to use the compassionate self/mind to engage with and work with specific problems such as anxiety, self-criticism, and shame trauma memory. Many ingredients of other therapies are incorporated here, especially CBT techniques although the therapy itself can be integrated well with other therapies.

expanded on in the following sections. CFT distinguishes between the therapy itself that involves the therapeutic relationship and the basic concepts and formulation processes, and the particular training in compassion with the individual. The specific training exercises of CFT, compassionate attending, thinking, behaviour, and imagery are part of what is called CMT.

Psycho-education

Our "tricky brains". The psycho-education process focuses on de-shaming and depersonalising, which focuses on the evolved brain and how it is a "tricky brain" (Gilbert, 2014). CFT recognises that our brains although intelligent, are essentially not "well designed" and are vulnerable to all sorts of problems, including damage (Gilbert, 1998, 2002). Central to this is that human brains have evolved cognitive capacities (what is termed the "new brain") to be able to imagine, anticipate, ruminate, have an objective sense of self, solve more complex problems, adapt more readily, create technology, etc. Essentially the evolved cortex and (primarily frontal lobes) play a crucial role in these complex cognitive capacities. However, there are a number of disadvantages to this "new brain" in that we can pursue harmful goals (e.g., be cruel, create terrorism and weapons of mass destruction). Furthermore, our capacities to reflect can stimulate anxiety, anger, fear (essentially our "old brain systems") and we can maintain these physiological systems in a state of activation in the body, which leads to physical and mental health difficulties. Alongside this we can monitor and judge ourselves, which can lead to many other types of problems including self-criticism, shame, narcissism, and self-harm. These types of issue are associated with mental health problems that regularly stimulate the threat processing systems (Gilbert, 2009).

In summary we have an evolved brain that is capable of significant dysfunction and is vulnerable to physical and mental health problems. Essential to the model is the view that "this is not our fault". In recognising that it is *not their fault*, it is also important for the client to recognise that if they just continue to let their attention get caught up in these cycles, they can do much harm to their own mental and/or physical health (as well as those around them). This is an important aspect of the therapy process, as the client understands that much of what goes on in their minds and lives is not their fault, but it is their responsibility to acknowledge this and work with it in order to make changes to alleviate the suffering of themselves and/or others.

In the context of ABI, additional information can be drawn into the psycho-education process for sense-making and de-shaming. Vulnerability to brain damage and the resulting neuropsychological impairments, especially to the "new brain" (although not excluding other areas) can help make sense of how a "tricky brain" can become even "trickier". Helping clients to see that their responses (e.g., disinhibition, emotion dysregulation) following ABI *are not their fault* is important for de-shaming and engagement. Emerging qualitative evidence highlights the importance of this additional neuropsychological information in the formulation and psycho-education process following ABI (Ashworth et al., 2015).

Our "tricky lives". Another central aspect of the psycho-education is recognising that we are shaped by our social contexts, which will influence who we are and who we will become. We are created by the integration of our genetics, our evolved brains, their functional nature, and our social circumstances and experiences. Central to this is that our brains have evolved to function in certain ways given our environmental and genetic influences, however, the evolved brain is sensitive to developing "different versions of itself" according to the social community that it is embedded in (Gilbert & Choden, 2013). Again this perspective can enable clients to have a useful insight into why they behave in ways they do in response to their evolved brain and life experiences, as well as recognising the possibility that this can be changed.

An example of this could be the client who has an ABI in the face of causing a road traffic collision due to him driving under the influence of alcohol. Through understanding and recognising that this individual is shaped by his social context, the client may begin to recognise that early adverse experiences lead him to abuse alcohol as a protective mechanism in response to his early experiences and fears. In recognising this, the client may begin to see that much of this was "not his fault" (but it is his responsibility), which can be helpful in de-shaming in order to bring about change.

The three systems of affect regulation. An important part of the model, which draws from multiple fields of study including neurophysiological and neuroscientific research with regards to emotional and psychological processing, is the recognition that we have different affect regulation systems that have evolved over time with different functions (Gilbert, 2000, 2005, 2010a,b). Specifically CFT suggests that we have three affect regulator systems. It is acknowledged that this understanding of systems is useful as a heuristic but it is still limited by the fact that there is much more complexity to these systems than we currently know. Figure 1 provides a diagrammatic representation of these systems, what is referred to as the "three circles model"

Figure 1: Three types of affect regulation system (taken from Gilbert (2009), with permission from Constable & Robinson).

within the actual therapy. Psychological well-being is associated with having balance in the stimulation and activation of these three affect regulator systems.

The threat–protection system

This system is designed to detect threat and respond accordingly for protection purposes (LeDoux, 1998). It is increasingly shown that this system has evolved to be dominant and we are therefore more likely to pay more attention to, process, and remember, negative than positive events more easily (Baumeister et al., 2001). This system is reasonably well understood in terms of its neurophysiology and learnt responses (e.g., classical conditioning and social contextual learning, Panksepp, 2010). As such, the threat system is fast to respond and will trigger emotions such as anger, anxiety, and disgust and activate certain defensive behaviours such as fight/flight, avoidance, and submission. This system can be activated by external threats (such as physical danger) as well as internal threats such as anger, anxiety, or internal fantasies (this is a key part of the philosophy of having a "tricky brain"). Those individuals who are highly shame-prone and self-critical are likely to have a dominating threat system to their internal and external worlds (Gilbert, 2009). Furthermore, even after a harmful event may have passed (e.g., being in a road traffic collision), further negative emotions can arise when the focus is on the harm or loss that has occurred (e.g., inability to be independent following ABI). Additionally, if a motivation is blocked (e.g., wanting to return to work but being unable to), this can also lead to stimulation of threat emotions (e.g., anxiety). Given the increased likelihood of the negative impact of the consequences of ABI, such as psychological distress, increased self-criticism, disruption to relationships, and increased social isolation such blocks will likely lead to threat system stimulation.

It is also important to note that heightened negative emotional states can exacerbate cognitive problems following ABI (e.g., Suhr & Gunstad, 2010), which consequently can lead to further distress and stimulation of the threat system. It could be argued that therapies that directly target up-regulation of a soothing response (which is fundamental to CFT) and down-regulation of a threat response may prove useful in reducing this exacerbation on cognitive problems. This is one of the ways in which CMT (described in the next section) may be particularly useful following ABI. Emerging evidence also indicates that CMT in the context of CFT following ABI can be useful (Ashworth et al., 2015).

In terms of neuroanatomical correlates, LeDoux's (1998) research on the two parallel systems implicated in threat processing is important, namely the subcortical "quick and dirty" route and the amygdalo-cortical route, where the role of the frontal lobes is to offer a more rationale cognitive processing response to threat related stimuli. Given that frontal lobe damage (affecting the amygdalo-cortical route) is commonplace following moderate to severe TBI and leads to emotion dysregulation (e.g., Damasio et al., 1990), damage to this circuitry means that the cognitive aspects of emotional processing of threat may be impaired. Furthermore, following ABI the use of the "quick and dirty" subcortical route may lead to actions with negative consequences (e.g., misperceiving threat, or overreacting to a threat).

The experience of self-criticism following ABI may well be effected by neuro-anatomical damage to the prefrontal cortex (PfC) and the dorsal anterior cingulate (dAC) given recent findings that the neural correlates of self-criticism are associated with these areas (Longe et al., 2010). These areas are also implicated in error processing, error monitoring, and resolution as well as inhibition that can be impaired following ABI.

The incentive–resource system

Positive emotions also play a role in threat appraisal and coping. This system is related to drive, excitement, vitality, and achievement. These activating positive emotions are associated with being driven to seek out and acquire the resources we need to survive. In their research, Depue and Morrone-Strupinsky (2005) outline this one form of positive emotion, which is stimulating and activating pleasure and excitement. This system is likely to be underpinned by neural circuitry and dopaminergic bodily response associated with pleasure and reward (e.g., Berridge & Kringelbach, 2013), which includes a range of meso-corticolimbic circuitry, both cortical and subcortical. When balanced with the two other systems this system guides us to important life goals. If there is an over reliance on achievement this can increase vulnerability to certain conditions such as depression; some dips in mood and depression are linked to pursuing achievement driven goals that are unobtainable, for example, people who cannot come to terms with losses such as injury (Gilbert, 1984).

Following ABI there may be a number of explanations for how the incentive–resource system may be negatively affected, with a few examples provided here. The loss of pleasurable activities following ABI may lead to a drop in the stimulation of this system. Conversely, continued attempts to pursue a pleasurable activity that one is no longer capable of doing can block the incentive–resource system and activate a threat system response. Following ABI some individuals may also attempt to focus on over activating this system if it is blunted, for example, by using illicit drugs to try to engage feelings of excitement and pleasure.

Once again, given the wide array of neural circuitry proposed to be involved in systems focused on achievement and incentive, it is likely that neural damage may affect this circuitry following ABI. Although this system does not play a central role in CFT per se, close attention needs to be given to how to enable stimulation of this system following ABI in a safe and useful way. ABI survivors may well require support to stimulate this system, including reducing detrimental drug use, returning to old pleasurable activities, and accessing new pursuits, which they find enjoyable and meaningful.

The affiliative and soothing–contentment system

This system is central to CFT. Research suggests that a specialised affect regulation system supports feelings of contentment, safeness, peacefulness, and affiliation. This sense of well-being links to particular types of positive affect (Depue & Morrone-Strupinsky, 2005; Mikulincer & Shaver, 2007) as well as neurophysiological stimulation of endorphins and oxytocin (Carter, 2014; Panksepp, 1998). This system is linked to activation of the parasympathetic nervous system (Porges, 2007). When this system is activated, there is a calming, resting, and contented state, where the individual is not threatened or seeking to achieve, thus indicating a calming of the threat–protection and incentive–resource systems. This sense of quiescence is associated with a sense of contentment, and is different from a relaxation response.

Of central importance to this system and CFT, is that this system is linked to affection and kindness. It is believed to have evolved with our attachment system, especially with the ability to recognise and respond to being cared for in a calm and contented way. Affection and kindness from others helps to soothe individuals and give them a sense of safeness when they are distressed. Essentially, CMT involves stimulation and training of this system, as it is essential to our sense of well-being. The type of caring we receive from others will stimulate the soothing–content system in different ways; specifically it is caring with warmth and affection (as opposed to this being absent) that is associated with feelings of contentment and soothing (MacDonald, 1992).

Oxytocin is implicated in relation to this system in various ways. For example, higher stress reactivity is linked with lower oxytocin levels and oxytocin activity is linked to social support and buffering stress (Heinrichs et al., 2003). Evidence also shows that oxytocin is implicated in the regulation of threat processing (e.g., Labuschagne et al., 2010). Individuals who have suffered trauma (such as abuse), early experiences of attachment difficulties, and threatening relationships can have difficulty in accessing this soothing system to regulate threat responses. Within CFT, this affect regulation system is viewed as inadequately available to individuals with high shame and self-criticism. Evidence to support this comes from studies that show that a relative inability to generate feelings of self-directed warmth, soothing, reassurance, and self-liking is common in those who are self-attacking (e.g., Gilbert, 2000; Gilbert et al., 2004).

Neuroscientific data enable us to have a greater understanding of the neural correlates underpinning nurturance, care, compassion, self-reassurance, and self-soothing. Studies suggest that compassionate responses and self-reassurance activate networks including the orbitofrontal cortex (OfC), ventrolateral PfC, insula, and subcortical areas, which have previously been linked to positive affect and affiliitation (Klimecki et al., 2013; Berridge & Kringelbach, 2013; Longe et al., 2010).

Following ABI, there are reasonable explanations to support a hypothesis that the soothing system may not be easily accessible. For example, those individuals with psychological distress, including self-criticism and shame may find it difficult to be as self-soothing, self-compassionate, and self-reassuring as shown in (non-brain injured) mental health populations. It is possible that there may be neurological and neurophysiological damage to the circuitry involved in the stimulation of this system. Although there are no specific studies on soothing abilities following ABI per se, it is possible to draw from studies that show that there are significant problems with regulation of emotion and behaviour following ABI (e.g., Dethier et al., 2013; Gainotti, 1993).

The experience of positive affect can elicit negative (threat system) reactions from some individuals; this may be as a result of traumatic early experiences and conditioned responses. Research is emerging to show that some individuals have a fear of compassion and this can make it challenging to engage in the psychotherapy process (Gilbert et al., 2012, 2013).

Given that the soothing–affiliative systems are linked with our attachment experiences, special consideration needs to be given to the :applicability and feasibility of CFT when working with individuals following ABI who have traumatic early histories combined with emotion dysregulation as a result of neurological damage. Conversely, it could be argued that CFT is well equipped to work with such individuals as the focus in CFT is to build up compassionate skills and capacities first (through CMT) in order to then work with psychological distress including self-criticism, shame, and other trauma related difficulties.

Within the literature on managing psychological distress post-ABI, there is a need for more research aimed specifically at targeting up-regulation of positive affect. The research has not tended to focus on the specific training of positive emotions such as soothing/content-ment that regulate threat and are central to CFT.

Formulation process

The formulation process in CFT, tied in with our "tricky lives" focuses on helping clients to understand how their early life experiences lead to coping (protective) strategies to defend against threat, as well as drive–excitement-based and affiliative soothing strategies and capaci-ties. These are explained in terms of being directed internally (how we relate to ourselves and regulate our own emotions) and externally (how we relate to others based on what we think is going on in their minds). With this understanding of the historical context (i.e., early experiences), consideration is given to what key fears may have arisen in response to these threats as well as the protective strategies used to protect against these key fears. Understanding how protective strate-gies may have had unforeseen consequences is also crucial. Following ABI, this formulation process helps us to understand a person's coping responses prior to injury and additionally, how the event and consequences of the injury interacts with and effects a person's coping strategies following ABI. The case study presented here provides an example of a CFT formulation integrating the consequences of the injury.

Compassionate mind training (CMT)

CMT focuses on developing abilities to generate feelings of self-reas-surance, self-soothing, and warmth, which can act as an antidote to a sense of threat. CMT involves a number of key abilities that include the *desire to care for the well-being* of another, *distress sensitivity* related to noticing and processing distress (rather than denial or dissociation), *sympathy*, *distress tolerance* for painful feelings "in another" rather than avoidance or attempting to control the other's emotions, *empathy* related to intuitive and cognitive abilities, and *non-judgment* related to the ability to be non-critical of the other's situation or behaviours (Gilbert & Procter, 2006). All these require the emotional tone of *warmth* (Gilbert & Procter, 2006). *Self*-compassion arises from exercising these capacities for self-to-self relating. Self-compassion can help reduce the sense of threat and create feelings of safeness. CMT involves training in specific exercises, that aim to activate the soothing system and cultivate compassion.

Before clients can begin to cultivate compassion, they need to be aware of where the mind habitually goes to, that is, the client needs to be able to be "mindful" of thoughts, feelings, and behaviours. Mindfulness and compassion should be seen as separate processes, which need to be specifically cultivated in their own ways (Gilbert & Choden, 2013). Through mindful awareness, the client can then learn to direct their focus of attention in ways that are helpful, for example, towards a compassionate mindset. Therefore mindfulness plays a role within CFT in the process of developing awareness of thoughts, feelings, and behaviours. However, central to this is the compassionate orientation that mindfulness is framed in.

The exercises focus on compassionate attending, thinking, behaviour, and imagery. Exercises (e.g., soothing rhythm breathing, compassionate safe place imagery, compassionate self imagery, compassionate letter writing) are introduced to clients and practised in sessions with reflection and exploration of their experiences. Clients practise these exercises outside of sessions.

CMT has a number of potential roles in neuro-rehabilitation settings. One key advantage of CMT is that the cognitive capacity required to engage in basic exercises is relatively minor and therefore will not likely put pressure on already compromised cognitive capacities. Given that CMT aims to activate parasympathetic nervous system activity, this can enable the client to increase their cognitive capacity (in relation to the notion that threat related affect can exacerbate cognitive impairments) via slowing down and pacing themselves in order to have mental space and time to compassionately address the challenges they are facing. To give an example, I worked with a client who had severe dysexecutive symptoms as well as emotional lability as a result of a moderate TBI. Despite repeated attempts to scaffold sessions and support use of a number of strategies, sessions were difficult to organise and engage with as the client jumped from topic to topic and became increasing labile. This was frustrating for the client, myself, and other rehabilitation professionals. CMT was introduced to the client as a way to calm the mind and this was used repeatedly at the start of, during and throughout psychotherapy sessions as well as other sessions across the neurorehabilitation setting. With regular practise, the client reported feeling more in control of her emotions and feeling more settled in the mind, thus providing a strategy to manage difficulties as well as giving space to work with other challenges she was facing. The staff (including myself) felt more able to engage with the client in a therapeutically beneficial way during the rehabilitations sessions.

Thus regular CMT practise may enable clients to feel less threat-focused emotions and feel more soothed in the context of facing challenges as a result of their ABI. Repeated practise of such exercises could have further implications for people with ABI, such as improved immune functioning and a reduced stress response. Given that CMT aims to develop capacities for compassion, of which mentalizing and empathy are elements, it is possible that CMT could be used to improve such difficulties (see section on *Social cognition deficits* below). However, until studies are conducted in this area it is unclear what benefits CMT can have for this population. Furthermore, CMT can support clients to develop soothing abilities to help them be able to engage with psychological pain and grief resulting from the losses and consequences of the ABI. A recent evaluation of CFT following ABI (Ashworth et al., 2015), which included CMT found that it improved well-being, helped clients to engage with their difficulties in a safe way, and provided new tools for coping with the consequences of the ABI. Crucially, the effectiveness of CMT for a variety of difficulties following ABI remains to be seen. However, it is clear that a certain level of cognitive ability and support may be required in order for individuals to be able to engage with and benefit from CMT.

Using compassion to engage with difficulties

As the client develops their compassion practices, if needed they can then be helped to use the compassionate self/mind to engage with and work with specific problems such as anxiety, depression, change in identity following ABI, self-criticism, grief, and shame based trauma memories to name a few. CFT draws from many different approaches in its application, including CBT techniques, although the therapy itself can be integrated well with other therapies.

In summary, although the background research and literature that CFT draws from is not necessarily new, it is the *drawing together and application* of this knowledge in a new and comprehensible way that can provide a useful and suitable approach in understanding and managing psychological distress following ABI. CFT is likely to be useful for those individuals who struggle with self-criticism and shame following ABI. Although there are no direct studies comparing CFT with CBT, it is likely that in the context of shame and self-criticism, CFT will likely prove more effective as it specifically addresses these processes and there is evidence to show poorer outcome following CBT with highly self-critical individuals. CFT can also provide clients with the tools to cope with the ongoing challenges of ABI in a way that addresses psychological distress via up-regulation of positive soothing based stimulation. Therapies such as CBT can be criticised for focusing only on down-regulation of negative processing, without a focus on up-regulation of positive affect. Additionally, CMT may provide especially useful for clients with reduced cognitive capacity, as the cognitive effort required to engage in basic exercises is relatively minor. In relation to this, it is also likely to be beneficial in neuro-rehabilitation settings where clients are struggling to engage with rehabilitation due to psychological distress. Anecdotally clients have also described how helpful the three affect regulation systems are (including the impact of neurological damage and neuropsychological deficits) in making sense of their psychological distress following ABI.

The emerging evidence base for CFT

Increasingly, CFT is being used to treat a wide range of mental health problems where there are shared transdiagnostic processes of shame and self-criticism. To date there are a number of studies using CFT for a range of disorders (e.g., depression, anxiety, personality disorder, eating disorders, psychosis), which have produced positive results (e.g., Gale et al., 2012; Gilbert & Irons, 2004; Goss & Allan, 2014; Judge et al., 2012; Laithwaite, 2010; Lucre & Corten, 2012; Mayhew & Gilbert, 2008). Although promising studies are predominantly clinical evaluation studies and case series with limitations, including small sample size and lack of a control group. A randomised controlled trial of the feasibility and efficacy of CFT intervention in psychosis reported positive outcomes (Braehler, et al., 2013).

Other compassionbased interventions are also beginning to show evidence for the positive benefits of compassion in promoting well-being, and positive mental health, namely the mindful self-compassion approach (Neff & Germer, 2013) and compassion cultivation training (Jazaieri, et al., 2013), which mostly focus on Buddhist roots and do not take an evolutionary perspective like CFT.

Experimental studies are beginning to provide some evidence for the positive impact of compassion-based training in physical and psychological well-being. Studies indicate that exposure to compassion training may affect activity in stress-relevant brain areas including the anterior cingulate and amygdala as well as reducing stress-induced immune and behavioural responses (e.g., Correa Mograbi, 2011; Lutz et al., 2008). When compared with a memory control group, compassion training leads to significant improvement in positive affect, even in response to others suffering (Klimecki et al., 2013), thus suggesting that compassion training can provide a new coping strategy for dealing with distress. These authors argue for functional neural plasticity as a result of compassion training, which activated a neural network including the medial orbitofrontal cortex, putamen, pallidum, and ventral tegmental area.

The dearth of research on CFT following ABI

To the author's knowledge there are only two published papers on the application of CFT following ABI and one published experimental study using a single session of compassionate imagery to try to improve empathy deficits following TBI (O'Neill & McMillan, 2012). The first paper briefly outlines the theoretical foundations of CFT and its application post-ABI, followed by a clinical case illustrating CFT for heightened self-criticism associated with mental health problems following TBI (Ashworth et al., 2011). This paper does not discuss the broader literature on self-criticism and shame that may be experienced by this population, nor is it a thorough application of CFT, as noted by the authors who highlight a core aspect of CFT that was not utilised in the intervention (the three affect regulation systems). The second is a case of depression associated with heightened self-criticism following stroke, which integrates a "third wave" approach, with some elements of CFT applied (Shields & Ownsworth, 2013). Although not a purely CFT approach, this case study provides a useful rationale for addressing self-criticism following stroke (Shields & Ownsworth, 2013). The O'Neill and McMillan (2012) study used a one-off compassionate imagery session aimed at improving empathy following ABI but is limited in its application as CMT is designed as a training process over time.

A more recent paper provides an evaluation of a combination of group and individual CFT (set within the context of holistic neuropsychological rehabilitation) for psychological distress following ABI including heightened self-criticism associated with symptoms of anxiety and depression (Ashworth et al., 2015). This evaluation provides both quantitative and qualitative data for twelve participants, which supports the use of CFT following ABI. Quantitative analysis provides evidence for a significant decrease in self-criticism, and symptoms of anxiety and depression. Statistical analysis also provides evidence for a significant increase in the ability to self-reassure. Qualitative themes indicate the utility of developing compas sion in combating self-criticism and creating a more helpful way of self-to-self and self-to-other relating in the context of ABI. This is the first group evaluation to provide some evidence for the utility of CFT following ABI, but is also limited by its small sample size (N = 12), lack of comparison group, and the likely role of multidisciplinary input in the change process. It is evident that further studies are required to draw conclusions as to the feasibility and effectiveness of CFT with this population. Prior to describing a case of CFT following ABI, special considerations for psychotherapy following ABI relevant to CFT are discussed.

Special considerations for CFT following ABI

Treatment of psychological distress post-ABI has followed trends from mainstream mental health. This presents some key limitations, especially given that the consequences of ABI can have unique aspects, which can differ from mainstream mental health difficulties. A fundamental issue is the nature of the organic damage, the neuropsychological sequelae, and the role this may play in the development and maintenance of mental health problems following ABI. ABI survivors who are aware of the changes in their cognitive ability (including difficulties with attention, comprehension, memory, and executive functioning) can be highly distressed by this. Psychological distress in the context of reduced cognitive capacity can further compromise functional performance. This can result in vicious cycles in which distress at cognitive failures takes up limited processing capacity and breeds further cognitive failures. At this point, traditional CBT interventions, which adopt a problem solving approach, can be difficult to access and engage with, given the inherent processing demands and executive functioning load. As a result, a person with an ABI, who presents with a complexity of interacting psychological and neuropsychological problems (especially cognitive problems), can struggle to benefit from a standard therapy, which may therefore require adaptation. Additionally psychological interventions following ABI have been critiqued for not closely or clearly considering the neuropsychological aspects of the injury as a key part of the *psychological* formulation (Ashworth et al., 2011). To address these limitations, it is essential to have a clear understanding of neuropsychological processes, as well as disruptions to these processes, in order to be able to implement, integrate, and adapt effective psychological interventions for individuals following ABI. This integration of approaches has recently been referred to as neuropsychotherapy and a useful definition comes from Kiiski-Maki (2013, p. 167):

> [neuropsychotherapy] is therapeutic work that combines methods eclectically from other fields of psychotherapy and, at the same time, modifies them according to the ABI child's [or adult's] special needs due to her [or his] unique set of neuropsychological symptoms.

Judd and Wilson (2005) describe some of the challenges encountered by clinicians providing psychotherapy to clients with ABI alongside specific modifications to practise. The paper outlines primarily neuropsychological (e.g., attention, memory problems), and psychological challenges (e.g., frustration, anger), and modifications to manage these difficulties. More recently Block and West (2013) provide special considerations for the implementation of psychotherapeutic interventions for individuals with TBI including cognitive, psychiatric/emotional, social factors as well as addressing difficulties with the therapeutic alliance. It is beyond the scope of this chapter to discuss all psychotherapeutic considerations in detail, however, the author recommends consideration of recent guidelines for neuro-psychotherapy practice with neurological patients (Laaksonen & Ranta, 2013). Two particularly pertinent neuro psychological issues that warrant further consideration here include anosognosia and social cognition deficits following ABI.

Anosognosia

Anosognosia, or lack of awareness of one's deficits, is a relatively common consequence of ABI. In contrast to the psychological process of denial, it is a neuropsychological deficit in self-awareness and is related to organic damage. It is essential to conduct an accurate assessment of the individual's impaired self-awareness. One should not confuse denial with a neuropsychological impairment in self-awareness, and vice versa as this can lead to counterproductive interventions (Langer & Padrone, 1992). Where there is complete anosognosia, the therapist needs to try to create a good therapeutic alliance by aiming to decrease the client's frustration through understanding their experience; Prigatano (1999) suggests that such patients will not present with resistance. Instead he suggests that resistance and denial will emerge in the patient with partial awareness (Prigatano, 1999, 2005). Increasing awareness should be a slow and careful process in order to enable the patient to recognise how their metacognitive functions have been affected in a safe way that does not lead to intolerable distress (Prigatano, 1999). For example, where there is partial awareness denial may play an important role in protecting the client from facing the reality of their problems prematurely.

It is likely (but not certain) that patients with complete anosognosia may not present with problematic levels of shame and self-criticism. However, evidence indicates that as awareness increases the patient's psychological distress can increase (Fleming et al., 1998). It is at this point that some individuals *may* be at risk of developing problematic levels of shame and/or self-criticism, as they may perceive the self as a failure/inadequate/damaged and be unable to emotionally tolerate this. In this context a proactive approach of developing a compassionate relationship with the self, when developing increased awareness of deficits, could be protective against the experience of further psychological distress as they may engage in a compassionate response towards difficulties and failures, rather than a self-critical response. Until more research has been conducted in this area, it is difficult to draw any conclusions. Case studies using CFT in the context of awareness deficits would be a useful first step in gaining an understanding of the feasibility of this approach with clients' with this type of problem.

Social cognition deficits

More recently there has been a trend to focus on common social cognition difficulties following ABI, which is of particular relevance in view of the relational focus within CFT (including self-to-self and self-to-other relational frames) as well as key elements of CMT, such as empathy. Studies to date indicate that following ABI, especially TBI, individuals are more likely to experience difficulties in social cognition including theory of mind (ToM), empathy, mentalizing, alexithymia, and emotional expression recognition difficulties (McDonald, 2013; McDonald & Flanagan, 2004; O'Neill & McMillan, 2012; Spikman et al., 2012; Williams & Wood, 2010). In a recent experimental study, O'Neill and McMillan (2012) attempted to reduce empathy deficits following severe TBI using one session of CMT. The results were not significant and the study is limited as CMT is based on the premise of building capacities for compassion through continued training. A comprehensive assessment and clear psychological and neuropsychological formulation is essential to understanding whether emotional factors (rather than solely cognitive) may play a central role in deficits such as empathy and mentalizing. For example, within the field of mental health, difficulties with mentalizing are common in those with personality disorders and it is believed that these problems may have an adaptive function, such as being key defensive or protective strategies in relation to early traumatic attachment experiences (e.g., Fonagy & Bateman, 2008). Thus the basis for empathy difficulties may in some cases be as a result of emotional factors as opposed to entirely cognitive deficits following ABI.

Social cognition deficits have a significant impact on relationships following ABI as the ability to communicate effectively can be disrupted (McDonald, 2013), for example, spousal relationship quality can be hampered especially in patients with frontal damage and associated cognitive deficits (Burridge et al., 2007). Such deficits can also interrupt the internal experience of the self, for example, when interoceptive ability is damaged. Self-awareness has been shown to be a necessary precondition for emotional empathy and the ability to separate one's own emotional experience from others (Decety & Meyer, 2008). Social cognition deficits including emotion expression recognition deficits are also found within mood disorders (Bombardier et al., 2010), although research in this area post-ABI is sparse. McDonald (2013) identifies multiple contributing factors to post-ABI social cognition deficits, including structural lesions underpinning emotional processing, cognitive deficits, mood disorders, and pre-existing personality traits. Few studies have specifically explored ways of ameliorating these deficits and evidence to date is limited by group heterogeneity, low power and weak group effects, and a small number of studies, primarily case studies. Yeates (2014) argues that for interventions to be true social cognition interventions, relational approaches need to be considered and as such presents evidence for an emotionfocused approach to couples therapy where one partner has both executive dysfunction and social cognition deficits (Yeates et al., 2013). While some individuals with ABI in experimental studies present with what appear to be fixed, absolute deficits in empathy and autonomic responses to social cues (e.g., Bechara et al., 1994), other tentative evidence indicates that following ABI such responses are not lacking, but can only be triggered at a higher threshold following ABI, thus being open to therapeutic manipulation (Evans et al., 2005; Yeates, 2014; Yeates et al., 2013). The current author's view aligns with the latter perspective and given that CFT aims to build or extend social cognition capacities such as empathy through CMT, this type of approach may have some promise with this population.

Given the challenges outlined above, it is clear that clinicians and researchers face a difficult task in identifying, implementing, and evidencing interventions aimed at reducing psychological distress and improving psychological well-being post-ABI. To demonstrate the application of CFT following ABI, the case of Mr A is presented.

Case example

Background

Mr A was a twenty-nine-year-old man when he was involved in a motor vehicle collision resulting in a severe TBI as well as orthopaedic and abdominal injuries. A computerised tomography (CT) scan revealed right inferior frontal lobe damage.

Neuropsychological assessment

Mr A had a number of cognitive consequences from his injury. Neuropsychological assessment revealed the following difficulties: slow speed of processing, poor selective attention, problems with encoding and retrieval of verbal material, and executive functioning difficulties. Assessment of executive functions specifically indicated problems with novel problem solving, completing unstructured tasks, and inhibition. Difficulties correlated with evidence for right frontal damage associated with disrupted inhibitory control and responses to distracters and retrieval problems (e.g., Aron et al., 2004).

Although Mr A was independent in basic daily tasks, such as self-care, he required support with more complex tasks such as managing bills and finances and organising his time. He was therefore provided with input from two support workers. In everyday life, Mr A was slower to take information, and had difficulties retrieving relevant information from his memory. He could be disinhibited at times (especially when he was frustrated) and struggled to complete more complex tasks without support, as well as problem solve in new situations. This left Mr A frustrated with his situation and feeling like a failure. However, Mr A's level of awareness of his difficulties was evident and the support workers who worked with him on a daily basis corroborated this perspective.

Reason for referral and psychological assessment

Mr A had previously received neuropsychological rehabilitation, however, he was re-referred due to experiencing severe depression following the break-up of his marriage. Psychological assessment revealed symptoms of depression and anxiety using the hospital anxiety and depression scale (HADS) (Zigmond & Snaith, 1983). Although the HADS has been found to be a useful measure of emotional distress following TBI, it is less reliable in predicting caseness in this population (Whelan-Goodinson et al., 2008). Clinical interview provided an in-depth assessment of Mr A's symptoms of anxiety and depression. Mr A reported that these symptoms were a significant change for him since the break-up of his marriage (further corroborated by his support workers), although he reported experiencing fatigue prior to the break-up, albeit not as severely. Given that Mr A's symptoms appeared directly after a stressful life event, and were not present prior to that, it was concluded that the psychological event rather than neurological damage was the precipitant to the onset of these changes. However, the neurological damage and resulting neuropsychological deficits were not ruled out as contributing factors and this was closely considered in relation to the intervention (see section on "Special considerations for CFT with Mr A" below). Mr A's depression symptoms included a lack of interest or pleasure in his day-to-day life, increased fatigue, difficulty sleeping, low motivation, low self-esteem, and suicidal ideation. Mr A's anxiety symptoms included worrying about the future, worrying about failure and getting things wrong, and physical sensations of anxiety.

Further difficulties included low self-efficacy (self efficacy scale; Schwarzer, & Jerusalem, 1995) and problematic levels of self-criticism with difficulties reassuring himself when things went wrong for him (forms of self-criticising and self reassurance scale (FSCRS); Gilbert et al., 2004). The FSCRS has robust psychometric properties, however, its accuracy within an ABI population has yet to be assessed. Preliminary pilot data using Rasch analysis indicates promising results. The FSCRS did not require adaptation for use with Mr A and the author has found it to be a useful clinical assessment tool for self-criticism and self-reassurance following ABI (Ashworth et al., 2012). See Table 2 for pre- and post-intervention measures. Mr A was highly avoidant of most activities and was struggling to be a father to his children, who both lived with his ex-wife, but visited him on weekends. Interview highlighted that Mr A related to himself in a highly critical way and he experienced significant shame associated with his inability to continue to fulfil the roles (including husband and father) and skills he felt he had prior to his injury. While he could retrospectively describe incidences of being self-critical, he found it difficult to monitor in the moment when he was being self-critical.

It was recommended that Mr A return for psychological therapy, specifically CFT, given the problematic self-criticism and shame and the impact it was having on his daily life and mental health. Mr A was provided with information on the content and process of CFT and he felt this way of working could help him. Despite his low mood, he felt

Table 2: Pre- and post-compassion focused therapy outcome measures

Measure	SD	$r_{xx\,a}$	Score		Interpretation		RCI_b	Reliable Change?
			Pre	Post	Pre	Post		
Inadequate (FSCRS)	8.44	0.90	34	9	Higher	Lower	7.40	Yes
Hated (FSCRS)	4.58	0.86	16	3	Higher	Lower	4.75	Yes
Reassure (FSCRS)	5.92	0.86	7	21	Lower	Higher	6.18	Yes
Anxiety (HADS)	4.0	0.92	12	7	Moderate	Normal	3.14	Yes
Depression (HADS)	3.9	0.88	19	7	Severe	Normal	3.74	Yes
Self-efficacy (SES)	4.2	0.76	20	33	Lower	Higher	5.70	Yes

Note: Forms of self-criticism and self-reassurance scale (FSCRS); hospital anxiety and depression scale (HADS); self-efficacy scale (SES); [a] Test–retest reliability of measure, [b] Jacobson & Truax (1991) formula for reliable change indices.

that he had some motivation to change. It is important to note that Mr A already knew the treating clinical psychologist (CP) as he had previously attended groups as part of holistic neuropsychological rehabilitation, which the therapist had been involved in facilitating. As a result of this there was already a good therapeutic alliance, which was an advantage for the process of psychotherapy. On reflection, without this therapeutic alliance in place at the start of therapy, this may have, at the very least, slowed down the process of change during the course of therapy.

Special considerations for CFT with Mr A

Given Mr A's neuropsychological profile, consideration was given to cognitive scaffolding in order to enable him to gain optimum benefit from the intervention. Mr A's session (sixty minutes) was divided into two slots of thirty minutes each with a break in between in order to reduce cognitive load placed on attentional capacities as well as to manage fatigue. The sessions were also split in terms of content, with the first thirty minutes focusing on formulation, barriers to compassion, reflections on specific examples of compassionate behaviours, especially with regard to behavioural experiments that had been set up and followed through on (essentially CFT). The second thirty minutes focused on CMT with time for reflection on session practise, followed by planning practise for the following week and looking ahead to consider if there were particular tasks/activities that may be more "threat" activating and incorporating compassionate practise and behaviour to take this into account. There was flexibility in session length and content depending on relevance and priority (but set within the sixty minute boundary).

The content of the sessions (especially those with psycho-education material) were delivered at a slower pace to suit Mr A. Regular reflection and summaries of discussions were provided initially in written format, although Mr A moved on to short audio-recordings on his smartphone for his "take home messages". Repeated reflections also supported encoding of memories and, where needed, Mr A was provided with cues to prompt retrieval from memory. With regards to addressing inhibition and problem solving difficulties, Mr A was re-introduced to the goal management framework (GMF; adapted from Robertson's (1996) *Goal Management Training*), which he had previously applied following rehabilitation but since being depressed had struggled to implement. With additional support from an occupational therapist working alongside Mr A and his support workers, he began to strategies such as the GMF in order to begin to plan activities and feel more able to organise his life. With regards to the CFT sessions, this enabled Mr A to "Stop and Think" before taking action (e.g., quickly getting caught up in frustration, moving off topic, or jumping ahead in session) as well as having a structured framework to think through problems, which supported the psychological work.

From a neurological perspective, Mr A's presentation of disinhibition and difficulties controlling his "threat" response correlated with damage to the amygdalo-cortical circuitry implicated in LeDoux's (1998) dual route threat processing framework. Given this, consideration was given to strategies that would interrupt processing via the "quick and dirty" route. The "Stop and Think" strategy described above was one such strategy, however, this was a cognitive strategy that did not directly down-regulate threat system processing. This provided a good rationale for the use of the CMT strategies with Mr A that directly aimed to activate the soothing system in order to down-regulate the threat system.

In relation to CMT, the exercises were recorded directly on to Mr A's smartphone in session and alerts were then set on his smartphone to remind him to practise. Furthermore, in addition to this Mr A received twice daily compassion text messages for four weeks sent via a pager (Martin-Saez et al., 2011) to support him in accessing his soothing system. These text messages were developed and agreed upon by Mr A so that they were personally meaningful; for example, "Hey Mr A, remember you have a tricky brain and that's not your fault". From a practical perspective, Mr A's smartphone was a key tool in supporting him to remember to practise and to actually complete the practise using the CMT recordings on his phone. One key aspect of the therapy and formulation was to consider the consequence of neuropsychological deficits in exacerbating and maintaining self-criticism and shame. This was clearly evident in Mr A's case, as he described experiences post-TBI where he had been unable to problem solve in the way he was able to prior to the injury and this resulted in failure experiences, which re-enforced self-criticism. Emerging evidence highlights the role of the right frontal lobe in the processing of self-criticism, and suggests that this is increased in those with heightened self-criticism. It was important to consider the role of neurological damage and impairments was integrated into the formulation process.

CFT psycho-education and formulation process

Feedback from previous ABI male service users who had received CFT indicated that the "three circles model" as an introduction to making sense of emotions and their regulation was exceptionally useful as a starting point. Therefore after review of the assessment findings including questionnaires and a reminder of what CFT involved, Mr A was introduced to the "three circles model" and collaboratively formulated his "three circles" in the context of brain injury. Alongside this the concept of the "tricky brain" and "it's not your fault" philosophy was shared and discussed and the recognition that much of what goes on in our minds and lives is "not our fault".

Within the context of ABI, and an extension of the CFT philosophy, is the emphasis and consideration given to the neuropsychological impact of the ABI on the brain, how a "tricky brain" can become even "trickier" as a result of neurological damage to various capacities and abilities. Mr A's consequences were drawn into the formulation process. In order to gain optimum benefits, Mr A was introduced to CMT early on in the therapy, in order to build up his soothing capacities, and provide more mental space for processing content within the sessions.

Using information from the assessment process, a collaborative formulation was developed and returned to. This initially focused on pre-injury history; understanding how Mr A's early life experiences, especially being bullied, led to fears of being rejected, of being seen as "rubbish" and of being discovered as not good enough. In response to these key fears Mr A developed a number of coping strategies to protect himself. Key external strategies included trying to please others, through putting their needs before his. Internal strategies involved setting exceptionally high standards for himself, expecting himself to be "at the top of his game" at all times. This "perfectionist ideal" was kept in check by using internal self-criticism as a function to defend against the fear of failure and rejection. It was further formulated that the way Mr A enabled himself to feel good about himself was by keeping his self-esteem high through focusing on overachievement. Discussion regarding affiliative soothing strategies indicated that this was not something that Mr A felt he could relate to historically. We revisited the "three circles model" from a historical (pre-injury) perspective and Mr A formulated that it was likely his systems were not well balanced and he had a tendency to use drive–excitement based strategies to manage fears of failure and rejection, and his soothing system was not particularly stimulated. The unforeseen consequences of Mr A's protective strategies were discussed; key outcomes of the strategy of pleasing others included Mr A feeling unfulfilled as his own goals and desires were not being pursued, and in turn Mr A would "try harder" to try to build feelings of success and satisfaction. Mr A would also be critical of himself in order to try to push himself, which he felt he had not recognised at the time, but this often resulted in excessively long hours at work prior to his injury. Mr A reflected that he did not feel that he struggled emotionally prior to his injury but he recognised that he likely had a vulnerability to self-criticism, which became a problem following the TBI.

The formulation process proceeded to focus on post-injury, specifically how the consequences of the TBI (including the break-up of his marriage) interacted with his pre-injury coping strategies. Mr A felt that he could no longer "please everyone" as he was not able to fulfil his roles as father, breadwinner, and husband to name a few. Mr A said that this was a key threat to himself in terms of his sense of identity as his experience of success was lost and his sense of being a failure and a rejection was activated. Mr A described feeling ashamed of these changes and unsurprisingly became even more self-critical, highlighting the evolution of the problematic self-criticism as a result of his perceived failures and losses. The unforeseen consequences of heightened self-criticism left Mr A feeling more ashamed and angry at himself, further leading him into a cycle of anxiety and depression.

Throughout this process, returning to the key philosophy of CFT, including drawing on the "tricky brain" and "tricky life" as not being Mr A's fault, was key in validating Mr A's experience. Central to this was experiencing the therapy as affiliative and supportive. These above processes were important in supporting Mr A to understand his problems and map them, which can lead to de-shaming and de-blaming.

Developing compassionate practice

Work with Mr A drew on the importance of compassion and affiliative emotions in relation to learning to stimulate the soothing system. Mr A was insightful about his self-criticism and its impact and was motivated to build a kinder, more compassionate self. Clearly, Mr A's awareness of his self-criticism and its negative impact was useful in being able to recognise and address it. Key to the process of CFT is the aim to develop compassionate social mentalities that come together through the desire to cultivate a compassionate self, a compassionate mind (Gilbert, 2009). Initially this focused on understanding and beginning CMT practices (such as the soothing rhythm breathing and safe place imagery exercises) to activate parasympathetic systems. This was practised both during and between sessions, with time to reflect in sessions as well as through completing a reflection sheet between sessions. The role of the therapist as a "compassion-focused" therapist is central in CFT, this means not only does the client experience the therapy in this context, but it also provides a social context to shape and model the cultivation of the compassionate mind.

Building a compassionate mind

Time was spent with Mr A exploring and developing a compassionate mind, considering what it would mean to have a greater sense of compassion as a person alongside CMT, specifically the "compassionate self" imagery exercise as well as a focus on what compassionate behaviours would incorporate. Behavioural experiments were set up to test out compassionate behaviours especially with regards to Mr A's role as a father. Mr A struggled to validate his children's difficult emotions and he was motivated to develop a more compassionate relationship with them. This compassionate perspective included recognising the challenges of being a parent and taking responsibility for this by committing to compassionate practice and parent training. At this stage, Mr A requested that his support workers receive training in understanding the model as he felt this would support him further. A two-hour session was spent collaboratively with Mr A and his workers sharing the philosophy of the model as well as planning and setting actions for what the support workers could do to enable Mr A to continue developing a compassionate sense of identity.

Using the compassionate mind to process trauma and threats

Time was spent in therapy reflecting on specific examples of behaviour change and the shift in Mr A's felt experience in relation to the activation of the soothing system. As Mr A began to feel the compassionate self was a key part of his sense of identity, some time was spent working with specific problems. Grief work around the loss of his pre-injury sense of identity was important. For example, Mr A felt much shame about the break-up of his relationship. Mr A engaged with this with a compassionate mentality and wrote a compassionate letter to himself about the break-up. Although very difficult to deal with, Mr A felt the process of working through this shifted the sense of shame and was replaced with a sense of sadness at the loss. Associated shame memories were also re-visited and re-scripted using a compassionate perspective. A particularly significant change occurred when Mr A was able to take a compassionate perspective of his ex-wife's situation, recognising that she, too, was "doing the best" she could in a tricky situation and this was "not her fault".

Cultivating the compassionate mind for the continued journey

Towards the end of therapy, Mr A began to take time to reflect on his own meaningful goals (i.e., not just focusing on others' needs) and what he would like to do. He had begun to engage in more social and occupational activities, instead of avoiding these situations. Throughout the therapy process, he had begun to increase his physical activity and this was one element that he had become more motivated to focus on further. He attended parent training to support his children and his motivation to be his compassionate self towards his children was reflected in his sharing of examples of the difference in the outcome of situations where he was "compassionate dad" compared with "critical dad". Final sessions focused on the ways in which Mr A wished to continuing building his compassionate self and how best to continue with the practises. Despite Mr A's cognitive difficulties, he was able to internalise the strategies and develop a compassionate mind. Mr A also reflected on feeling that his cognitive capacities were "better and clearer" when he was in a compassionate mind *vs.* a threatened mind. The context of life in the future was discussed, recognising that life would continue to be tricky, more so at some times than others. It was clear that Mr A was motivated to move forwards with a compassionate sense of identity, and he was opening up to dealing with the future with a compassionate perspective as well as giving capacity to his achievement system. This was further supported by input from the occupational therapist and his support workers.

Outcome

Mr A's behaviour change was evident through the process of therapy and was evident in both his increased sense of well-being and happiness. Post intervention measures reflected Mr A's positive changes, showing a significant reduction in self-criticism, anxiety, and depression as well as an increase in the ability to reassure/soothe the self.

Mr A began to build more social connections within his community, organising regular poker nights with help from his support workers. He also began to take more of an autonomous role in his life, doing more tasks at home such as gardening and spending more time with his children. Although Mr A was unable to return to work, he began to fill his time with more meaningful activities and planned to walk a marathon in aid of a local brain injury charity. Although he still required some minimal support, one significant change was reduction in the amount of time he required from support workers. For further information on Mr A's personal reflections and changes as a result of CFT see his description of the process in Ashworth and Wright (2013).

Conclusion

In summary, there is emerging evidence to demonstrate that self-criticism and shame are processes that may be contributing to and maintaining psychological distress following ABI. Given this, it is recommended that psychological assessment should include a specific focus on identifying those prone to shame and self-criticism following ABI. Research aimed at increasing our understanding of self-criticism and shame and the interacting role of neuropsychological factors in these processes following ABI is necessary. As CFT aims to ameliorate self-criticism and shame through developing greater compassion and the evidence for its effectiveness in mainstream mental health is emerging, it seems logical that CFT may be a useful intervention in reducing self-criticism and shame associated with psychological distress following ABI, although special considerations need to be given in the context of neurological damage and neuropsychological impairments. Clinicians and researchers considering implementing CFT following ABI, should proceed with these considerations and due caution. Certainly the case example presented highlights its effectiveness for this specific individual. However, the dearth of research in this area and the limitations of existing studies, restricts interpretation and generalisation of findings. Research aimed at the understanding the potential role for CMT in the neuro-rehabilitation context is warranted. Another area that lacks any published studies following ABI, is the experience of fear of positive emotions including self-compassion, which can be blocks to the change process. While CFT seems to have promise for this population, empirical research regarding the effectiveness of CFT following ABI, is needed in order to understand its true potential (Wilson, 2013).

Acknowledgements

I am grateful to two reviewers for their comments, and I am thankful for Clara Murray's support and guidance in the preparation of this manuscript. I am especially grateful to Mr A for permission to write about his CFT experience, as well as his feedback on this manuscript. Thanks to Wes as always for his patience.

References

Anson, K., & Ponsford, J. (2006). Coping and emotional adjustment following traumatic brain injury. *Journal of Head Trauma and Rehabilitation, 21*: 248–259.

Aron, A. R., Robbins, T. W., & Poldrack, R. A. (2004). Inhibition and the right frontal inferior cortex. *Trends in Cognitive Science, 8*: 170–177.

Ashworth, F., & Wright, A. (2013). Adrian's story: dealing with the potholes along the way. In: B. A. Wilson, J. Winegardner, & F. Ashworth (Eds.), *Life After Brain Injury: Survivors' Stories* (pp. 131–142). Hove: Psychology Press.

Ashworth, F., Bauch, E., & Bateman, A. (2012). A pilot analysis of the forms of self-criticism and self-reassurance scale in acquired brain injury. *Brain Injury, 26*(4–5): 752.

Ashworth, F., Clarke, A., Jones, L., Jennings, C., & Longworth, C. (2015). Compassion focused therapy following acquired brain injury. *Psychology and Psychotherapy: Theory, Research and Practice. 88*(2): 143–162.

Ashworth, F., Gracey, F., & Gilbert, P. (2011). Compassion focused therapy after traumatic brain injury: theoretical foundations and a case illustration. *Brain Impairment, 12*: 128–139.

Baumeister, R. F., Bratslavsky, E., Finkenauer, C., & Vohs, K. D. (2001). Bad is stronger than good. *Review of General Psychology, 5*(4): 323–370.

Bechara, A., Damasio, A. R., Damasio, H., & Anderson, S. W. (1994). Insensitivity to future consequences following damage to human prefrontal cortex. *Cognition, 50*: 7–15.

Bedard, M., Felteau, M., Marshall, S., Cullen, N., Gibbons, C., Dubois, S., Maxwell, H., Mazmanian, D., Weaver, B., Rees, L., Gainer, R., Klein, R., & Moustgaard, A. (2013). Mindfulness-based cognitive therapy reduces symptoms of depression in people with a traumatic brain injury: results from a randomised controlled trial. *Journal of Head Trauma and Rehabilitation*.

Berridge, K. C., & Kringelbach, M. L. (2013). Neuroscience of affect: brain mechanisms of pleasure and displeasure. *Current Opinion in Neurobiology, 23*(3): 294–303.

Block, C. K., & West, S. E. (2013). Psychotherapeutic treatment of survivors of traumatic brain injury: review of the literature and special considerations. *Brain Injury, 2*: 775–788. doi: 10.3109/02699052.2013.775487.

Bombardier, C. H., Fann, J. R., Temkin, N. R., Esselman, P. C., Barber, J., & Dikmen, S. S. (2010). Rates of major depressive disorder and clinical outcomes following traumatic brain injury. *Journal of the American Medical Association, 303*(19): 1938–1945.

Bradbury, C. L., Christensen, B. K., Lau, M. A., Ruttan, L. A., Arundine, A. L., & Green, R. E. A. (2008). The efficacy of cognitive behaviour therapy in the treatment of emotional distress after acquired brain injury. *Archives of Physical Medicine and Rehabilitation, 89*(12): S61–S68.

Braehler, C., Gumley, A., Harper, J., Wallace, S., Norrie, J., & Gilbert, P. (2013). Exploring change processes in compassion focused therapy in psychosis: results of a feasibility randomized controlled trial. *British Journal of Clinical Psychology, 52*(2): 199–214.

Burridge, A., Williams, W. H., Yates, P., Harris, A., & Ward, C. (2007). Spousal relationship satisfaction following acquired brain injury: the role of insight and socio-economic skill. *Neuropsychological Rehabilitation, 17*(1): 95–105.

Campos, R. C., Besser, A., Ferreira, R., & Blatt, S. J. (2012). Self-criticism, neediness, and distress among women undergoing treatment for breast cancer: a preliminary test of the moderating role of adjustment to illness. *International Journal of Stress Management, 19*(2): 151–174.

Carter, S. C. (2014). Oxytocin pathways and the evolution of human behavior. *Annual Review of Psychology, 65*: 17–39.

Correa Mograbi, G. J. (2011). Meditation and the brain: attention control and emotion. *Mens Sana Monographs, 9*(1): 276–283.

Cox, B. J., MacPherson, P. S. R., Enns, M. W., & McWilliams, L. A. (2004). Neuroticism and self-criticism associated with posttraumatic stress disorder in a nationally representative sample. *Behavior Research and Therapy, 42*: 105–114.

Curran, C. A., Ponsford, J. L., & Crowe, S. (2000). Coping strategies and emotional outcome following traumatic brain injury: a comparison with orthopedic patients. *The Journal of Head Trauma Rehabilitation, 15*(6): 1256–1274.

Dalai Lama (1995). *The Power of Compassion*. Delhi: HarperCollins.

Damasio, A. R., Tranel, D., & Damasio, H. (1990). Individuals with socio-pathic behaviour caused by frontal lobe damage fail to respond auto-matically to social stimuli. *Behavioural Brain Research*, *41*(2): 81–94.

Decety, J., & Meyer, M. (2008). From emotion resonance to empathic under-standing: a social developmental neuroscience account. *Development & Psychopathology*, *20*: 1053–1080.

Depue, R. A., & Morrone-Strupinsky, J. V. (2005). A neurobehavioural model of affiliative bonding. *Behavioural and Brain Sciences*, *28*: 313–395.

Dethier, M., Blairy, S., Rosenberg, H., & McDonald, S. (2013). Emotional regulation impairments following severe traumatic brain injury: an investigation of the body and facial feedback effects. *Journal of International Neuropsychological Society*, *19*(4): 367–379.

Dowswell, G., Lawler, J., Dowswell, T., Young, J., Forster, A., & Hearn, J. (2000). Investigating recovery from stroke: a qualitative study. *Journal of Clinical Nursing*, *9*(4): 507–515.

Dozois, D. J. A., Seeds, P. M., & Collins, K. A. (2009). Transdiagnostic approaches to the prevention of depression and anxiety. *Journal of Cognitive Psychotherapy*, *23*: 44–59.

Dunkley, D. M., Sanislow, C. A., Grilo, C. M., & McGlashan, T. H. (2009). Self-criticism versus neuroticism in predicting depression and psychosocial impairment for 4 years in a clinical sample. *Comprehensive Psychiatry*, *50*(4): 335–346.

Evans, C. E. Y., Bowman, C. H., & Turnbull, O. H. (2005). Subjective awareness on the Iowa Gambling Task: the key role of emotional expe-riences schizophrenia. *Journal of Clinical and Experimental Neuro-physchology*, *27*(6): 656–664.

Fennig, S., Hadas, A., Itzhaky, L., Roe, D., Apter, A., & Shahar, G. (2008). Self-criticism is a key predictor of eating disorder dimensions among impatient adolescent females. *The International Journal of Eating Disorders*, *41*: 762–765.

Fleming, J. M., Strong, J., & Ashton, R. (1998). Cluster analysis of self-awareness levels in adults with traumatic brain injury and relationship to outcome. *Journal of Head Trauma Rehabilitation*, *13*(5): 39–51.

Fonagy, P., & Bateman, A. (2008). The development of borderline person-ality disorder—a mentalizing model. *Journal of Personality Disorders*, *22*(1): 4–21.

Freeman, A., Adams, M., & Ashworth, F. (2014). An exploration of the experience of self in the social world for men following traumatic brain injury. *Neuropsychological Rehabilitation*, *25*: 189–215.

Gainotti, G. (1993). Emotional and psychosocial problems after brain injury. *Neuropsychological Rehabilitation*, *3*(3): 259–277.

Gale, C., Gilbert, P., Read, N., & Goss, K. (2012). An evaluation of the impact of introducing compassion focused therapy to a standard treatment programme for people with eating disorders. *Clinical Psychology and Psychotherapy, 21*: 1–12. doi: 10.1002/cpp.1806

Germer, C. K., & Siegel, R. D. (Eds.) (2012). *Wisdom and Compassion in Psychotherapy: Deepening Mindfulness in Clinical Practice*. Guilford: Guildford Press.

Gilbert, P. (1984). *Depression: From Psychology to Brain State*. London: Lawrence Erlbaum.

Gilbert, P. (1997). The evolution of social attractiveness and its role in shame, humiliation, guilt and therapy. *British Journal of Medical Psychology, 70*: 113–147.

Gilbert, P. (1998). The evolved basis and adaptive functions of cognitive distortions. *British Journal of Medical Psychology, 71*: 447–464.

Gilbert, P. (2000). Social mentalities: internal "social" conflicts and the role of inner warmth and compassion in cognitive therapy. In: P. Gilbert & K. G. Bailey (Eds.), *Genes on the Couch: Explorations in Evolutionary Psychotherapy* (pp. 118–150). Hove: Brunner-Routledge.

Gilbert, P. (2002). Evolutionary approaches to psychopathology and cognitive therapy. *Special Edition: Evolutionary Psychology and Cognitive Therapy, Cognitive Psychotherapy: An International Quarterly, 16*: 263–294.

Gilbert, P. (2005). *Compassion: Conceptualisations, Research and Use in Psychotherapy*. Hove: Routledge.

Gilbert, P. (2006). Evolution and depression: issues and implications. *Psychological Medicine, 36*: 287–297.

Gilbert, P. (2007). The evolution of shame as a marker of relationship security. In: J. L. Tracey, R. W. Robins, & J. P. Tangney (Eds.), *The Self-Conscious Emotions: Theory and Research* (pp. 283–309). New York: Guilford Press.

Gilbert, P. (2009). *The Compassionate Mind: A New Approach to the Challenges of Life*. London: Constable & Robinson.

Gilbert, P. (2010a). *Compassion Focused Therapy: Distinctive Features*. London: Routledge.

Gilbert P. (Ed.) (2010b). Compassion focused therapy. *Special Issue: International Journal of Cognitive Therapy, 3*(2): 95–210.

Gilbert, P. (2014). The origins and nature of compassion focused therapy. *British Journal of Clinical Psychology, 53*(1): 6–41. doi: 10.1111/bjc.12043.

Gilbert, P., & Choden (2013). *Mindful Compassion*. London: Robinson.

Gilbert, P., & Irons, C. (2004). A pilot exploration of the use of compassionate imagery in a group of self critical people. *Memory, 12*(4): 507–516.

Gilbert, P., & Irons, C. (2005). Focused therapies and compassionate mind training for shame and self-attacking. In: P. Gilbert (Ed.), *Compassion: Conceptualisations, Research and Use in Psychotherapy*. (pp. 263–325). Hove: Routledge.

Gilbert, P., & Miles, J. N. (2000). Sensitivity to social put-down: its relationship to perceptions of social rank, shame, social anxiety, depression, anger and self-other blame. *Personality and Individual Differences, 29*(4): 757–774.

Gilbert, P., & Procter, S. (2006). Compassionate mind training for people with high shame and self-criticism: overview and pilot study of a group therapy approach. *Clinical Psychology & Psychotherapy, 13*(6): 353–379.

Gilbert, P., Clark, M., Hempel, S., Miles, J. N. V., & Irons, C. (2004). Criticising and reassuring oneself: an exploration of forms, styles and reasons in female students. *British Journal of Clinical Psychology, 43*: 31–50.

Gilbert, P., McEwan, K., Catarino, F., Baião, R., & Palmeira, L. (2013). Fears of happiness and compassion in relationship with depression, alexithymia, and attachment security in a depressed sample. *British Journal of Clinical Psychology, 53*(2): 228–244.

Gilbert, P., McEwan, K., Gibbons, L., Chotai, S., Duarte, J., & Matos, M. (2012). Fear of compassion and happiness in relation to alexithymia, mindfulness and self-criticism. *Psychology and Psychotherapy, 85*(4): 374–390.

Goss, K., & Allan, S. (2014). The development and application of compassion-focused therapy for eating disorders (CFT-E). *British Journal of Clinical Psychology*, I>53*(1): 62–77.

Goss, K., Gilbert, P., & Allan, S. (1994). An exploration of shame measures. I: the "other as shamer scale". *Personality and Individual Differences, 17*: 713–717.

Hagger, B. F. (2011). An exploration of self-disclosure after traumatic brain injury. *Thesis submitted to the University of Birmingham for the degree of Doctor of Philosophy*. Unpublished manuscript.

Harvey, A., Watkins, E., Mansell, W., & Shafran, R. (2004). *Cognitive Behavioural Processes Across Psychological Disorders: a Transdiagnostic Approach to Research and Treatment*. Oxford: Oxford University Press.

Heinrichs, M., Baumgartner, T., Kirschbaum, C., & Ehlert, U. (2003). Social support and oxytocin interact to suppress cortisol and subjective responses to psychosocial stress. *Biological Psychiatry, 54*(12): 1389–1398.

Hibbard, M. R., Uysal, S., Kepler, K., Bogdany, J., & Silver, J. (1998). Axis I psychopathology in individuals with traumatic brain injury. *Journal of Head Trauma and Rehabilitation, 13*(4): 24–39.

Hodgson, J., McDonald, S., Tate, R., & Gertler, P. (2005). A randomised controlled trial of a cognitive-behavioural therapy program for managing social anxiety after acquired brain injury. *Brain Impairment, 6*(3): 169–180.

Hsieh, M. Y., Ponsford , J., Wong, D., Schönberger, M., Taffe, J., & McKay, A. (2012). Motivational interviewing and cognitive behaviour therapy for anxiety following traumatic brain injury: a pilot randomised controlled trial. *Neuropsychological Rehabilitation, 22*(4): 585–608.

Hutton, P., Kelly, J., Lowens, I., Taylor, P., & Tai, S. (2013). Self-attacking and self-reassurance in persecutory delusions: a comparison of healthy, depressed and paranoid individuals. *Psychiatry Research, 205*: 127–136.

Jacobson, N. S., & Truax, P. (1991). Clinical significance: a statistical approach to defining meaningful change in psychotherapy research. *Journal of Consulting and Clinical Psychology, 59*: 12–19.

Jazaieri, H., Jinpa, G. T., McGonigal, K., Rosenberg, E. L., Finkelstein, J., Simon-Thomas, E., Cullen, M., Doty, J. R., Gross, J. J., & Goldin, P. R. (2013). Enhancing compassion: a randomized controlled trial of a compassion cultivation training program. *Journal of Happiness Studies, 14*(4): 1113–1126.

Jones, L., & Morris, R. (2013). Experiences of adult stroke survivors and their parent carers: a qualitative study. *Clinical Rehabilitation, 27*(3): 272–280.

Judd, D., & Wilson, S. L. (2005). Psychotherapy with brain injury survivors: an investigation of the challenges encountered by clinicians and their modifications to therapeutic practice. *Brain Injury, 19*: 437–449.

Judge, L., Cleghorn, A., McEwan, K., & Gilbert, P. (2012). An exploration of group-based compassion focused therapy for a heterogeneous range of clients presenting to a community mental health team. *International Journal of Cognitive Therapy, 5*: 420–429. doi:10.1521/ijct.2012.5.4.420

Kiiski-Maki, H. (2013). Helping children with acquired brain injury to engage in a neuropsychotherapeutic process. In: R. Laasksonen & M. Ranta (Eds.), *Introduction to Neuropsychotherapy. Guidelines for Rehabilitation of Neurological and Neuropsychiatric Patients Throughout the Lifespan* (pp. 143–170). Hove: Psychology Press.

King, N. S. (2002). Perseveration of traumatic memories in PTSD: a cautionary note regarding exposure based psychological treatments for PTSD when head injury and dysexecutive impairment are also present. *Brain Injury, 16*(1): 65–74.

Klimecki, O. M., Leibergh, S., Lamm, C., & Singer, T. (2013). Functional neural plasticity and associated changes in positive affect after compassion training. *Cerebral Cortex, 23*(7): 1552–1561.

Laaksonen, R., & Ranta, M. (2013). Introduction to Neuropsychotherapy. *Guidelines for Rehabilitation of Neurological and Neuropsychiatric Patients Throughout the Lifespan.* Hove: Psychology Press.

Labuschagne, I., Phan, K. L., Wood, A., Angstadt, M., Chua, P., Heinrichs, M., Stout, J. C., & Nathan, P. J. (2010). Oxytocin attenuates amygdala reactivity to fear in generalized social anxiety disorder. *Neuropsychopharmacology, 35*(12): 2403–2413.

Laithwaite, H. M. (2010). Recovery after psychosis: a compassion focused recovery approach to psychosis in a forensic mental health setting. *Unpublished doctoral thesis,* University of Glasgow.

Langer, K. G., & Padrone, F. J. (1992). Psychotherapeutic treatment of awareness in acute rehabilitation of traumatic brain injury. *Neuropsychological Rehabilitation, 1*: 59–70.

LeDoux, J. (1998). *The Emotional Brain.* London: Weidenfeld & Nicolson.

Longe, O., Maratos, F. A., Gilbert, P., Evans, G., Volker, F., Rockliff, H., & Rippon, G. (2010). Having a word with yourself: neural correlates of self-criticism and self-reassurance. *Neuroimage, 49*(2): 1849–1856.

Lucre, K. M., & Corten, N. (2012). An exploration of group compassion-focused therapy for personality disorder. *Psychology and Psychotherapy: Theory, Research and Practice, 86*(4): 387–400. doi: 10.1111/j.2044-8341.2012.02068.x

Lutz, A., Brefczynski-Lewis, J., Johnstone, T., & Davidson, R. J. (2008). Regulation of the neural circuitry of emotion by compassion meditation: effects of the meditative expertise. *Public Library of Science, 3*: 1–5.

Luyten, P., Sabbe, B., Blatt, S. J., Meganck, S., Jansen, B., De Grave, C., Maes, F., & Corveleyn, J. (2007). Dependency and self-criticism: relationship with major depressive disorder, severity of depression, and clinical presentation. *Depression and Anxiety, 24*(8): 586–596.

MacBeth, A., & Gumley, A. (2012). Exploring compassion: a meta-analysis of the association between self-compassion and psychopathology. *Clinical Psychology Review, 32*: 545–552.

MacDonald, K. (1992). Warmth as a developmental construct: an evolutionary analysis. *Child Development, 63*(4): 753–773.

Martin-Saez, M., Deakins, J., Winson, R., Watson, P., & Wilson, B. A. (2011). A 10 year follow up of a paging service for people with memory and planning problems within a healthcare system: How do recent users differ from the original users? *Neuropsychological Rehabilitation, 21*(6): 769–778.

Mayhew, S. L., & Gilbert, P. (2008). Compassionate mind training with people who hear malevolent voices: a case series report. *Clinical Psychology & Psychotherapy, 15*: 113–138.

McDonald, S. (2013). Impairments in social cognition following severe traumatic brain injury. Journal of the International *Neuropsychological Society, 19*: 231–246. doi:10.1017/S1355617712001506

McDonald, S., & Flanagan, S. (2004). Social perception deficits after traumatic brain injury: the interaction between emotion recognition, mentalising ability and social communication. *Neuropsychology, 18*: 572–579.

McEvoy, P. M., Nathan, P., & Norton, P. J. (2009). Efficacy of transdiagnostic treatments: a review of published outcome studies and future research directions. *Journal of Cognitive Psychotherapy, 23*: 27–40.

Mikulincer, M., & Shaver, P. R. (2007). Boosting attachment security to promote mental health, prosocial values, and inter-group tolerance. *Psychological Inquiry, 18*(3): 139–156.

Miron, L. R. (2013). A comparison of self-compassion and mindfulness in predicting psychological distress, health status, and well-being. Doctoral Dissertation, Northern Illinois University.

Neff, K. D., & Germer, C. K. (2013). A pilot study and randomised controlled trial of the mindful self-compassion program. *Journal of Clinical Psychology, 69*(1): 28–44.

Nilsson, I., Jansson, L., & Norberg, A. (1997). To meet with a stroke: patients' experiences and aspects seen through a screen of crises. *Journal of Advanced Nursing, 25*(5): 953–963.

O'Neill, M., & McMillan, T. M. (2012). Can deficits in empathy after head injury be improved by compassionate imagery? *Neuropsychological Rehabilitation, 22*: 836–851.

Ownsworth, T. L., & Oei, T. P. (1998). Depression after traumatic brain injury: conceptualization and treatment considerations. *Brain Injury, 12*(9): 735–751.

Panksepp, J. (1998). The periconscious substrates of consciousness: affective states and the evolutionary origins of the SELF. *Journal of Consciousness Studies, 5*(5–6): 566–582.

Panksepp, J. (2010). Affective neuroscience of then emotional BrainMind: evolutionary perspectives and implications for understanding depression. *Dialogues in Clinical Neuroscience, 12*(4): 533.

Porges, S. W. (2007). The polyvagal perspective. *Biological Psychology, 74*(2): 116–143.

Priel, B., & Shahar, G. (2000). Dependency, self-criticism, social context and distress: comparing moderating and mediating models. *Personality and Individual Differences, 28*(3): 515–525.

Prigatano, G. P. (1999). *Principles of Neuropsychological Rehabilitation*. New York: Oxford University Press.

Prigatano, G. P. (2005). Disturbances of self-awareness and rehabilitation of patients with traumatic brain injury: a 20-year perspective. *Journal of Head Trauma Rehabilitation, 20*: 19–29.

Rasquin, S. M. C., van de Sande, P., Praamstra, A. J., & van Heugten, C. M. (2009). Cognitive-behavioural intervention for depression after stroke: five single case studies on effects and feasibility. *Neuropsychological Rehabilitation, 19*: 208–222.

Rector, N. A., Bagby, R. M., Segal, Z. V., Joffe, R. T., & Levitt, A. (2000). Self-criticism and dependency in depressed patients treated with cognitive therapy or pharmacotherapy. *Cognitive Therapy and Research, 24*: 571–584.

Robertson, I. H. (1996). *Goal Management Training: A Clinical Manual*. Cambridge, UK: PsyConsult.

Robinson, R. G., & Spalletta, G. (2010). Poststroke depression: a review. *Canadian Journal of Psychiatry, 55*(6): 341–349.

Schwarzer, R., & Jerusalem, M. (1995). Generalized self-efficacy scale. Measures in Health Psychology: a User' Portfolio. *Causal and Control Beliefs, 1*: 35–37.

Seel, R. T., & Kreutzer, J. S. (2003). Depression assessment after traumatic brain injury: an empirically based classification method. *Archives of Physical Medicine and Rehabilitation, 84*: 1621–628.

Shields, C., & Ownsworth, T. (2013). An integration of third wave cognitive behavioural interventions following stroke: a case study. *Neuro-Disability & Psychotherapy, 1*(1): 39–69.

Singer, T., & Lamm, C. (2009). The social neuroscience of empathy. *Annals of the New York Academy of Sciences, 1156*(1): 81–96.

Soo, C., & Tate, R. (2007). Psychological treatment for anxiety in people with traumatic brain injury. *Cochrane Database of Systematic Reviews, 3*, CD005239.

Spikman, J. M., Timmerman, M. E., Milders, M. V., Veenstra, W. S., & van der Naalt, J. (2012). Social cognition impairments in relation to general cognitive deficits, injury severity, and prefrontal lesions in traumatic brain injury patients. *Journal of Neurotrauma, 29*(1): 101–111.

Suhr, J. A., & Gunstad, J. (2010). "Diagnosis threat": the effect of negative expectations on cognitive performance in head injury. *Journal of Clinical and Experimental Neuropsychology, 24*(4): 448–457.

Tangney, J. P., & Dearing, R. (2002). *Shame and Guilt*. New York: Guilford.

Van Dam, N. T., Sheppard, S. C., Forsyth, J. P., & Earleywine, M. (2011). Self-compassion is a better predictor than mindfulness of symptom severity and quality of life in mixed anxiety and depression. *Journal of Anxiety Disorders*, 25(1): 123–130.

Waldron, B., Casserly, L. M., & O'Sullivan, C. (2013). Cognitive behavioural therapy for depression and anxiety in adults with acquired brain injury. What works for whom? *Neuropsychological Rehabilitation*, 23(1): 64–101.

Whelan-Goodinson, R., Ponsford, J., & Schonberger, M. (2008). Association between psychiatric state and outcome following traumatic brain injury. *Journal of Rehabilitation Medicine*, 40(10): 850–857.

Williams, C., & Wood, R. L. (2010). Alexithymia and emotional empathy following traumatic brain injury. *Journal of Experimental Clinical Neuropsychology*, 32(3): 259–267.

Wilson, B. A. (2013). Neuropsychological rehabilitation: state of the science. *South African Journal of Psychology*, 43(3): 267–277.

Yeates, G. N. (2014). Social cognition interventions in neuro-rehabilitation: an overview. *Advances in Clinical Neuroscience and Rehabilitation*, 14(1): 12–13.

Yeates, G. N., Edwards, A., Murray, C., & Creamer, N. (2013). The use of emotionally-focused couples therapy (EFT) for survivors of acquired brain injury with social cognition and executive functioning impairments and their partners: a case series analysis. *Neuro-Disability & Psychotherapy*, 1(2): 151–194.

Zigmond, A. S., & Snaith, R. P. (1983). The hospital anxiety and depression scale. *Acta Psychiatrica Scandinavica*, 67(6): 361–370.

Zuroff, D. C., Santor, D. A., & Mongrain, M. (2004). Dependency, self criticism, and maladjustment. In: J. S. Auerbach, K. J. Levy, & C. E. Schaffer (Eds.), *Relatedness, Self-definition and Mental Representation: Essays in Honor of Sidney J. Blatt* (pp. 75–90). London: Brunner-Routledge.

CHAPTER FIVE

Facing degeneration with compassion on your side: using compassion focused therapy with people with a diagnosis of a dementia*

Rebecca Poz

Background

The Prime Minister's dementia challenge, in the UK, gave a seal of approval to the early diagnosis of dementia, that in itself is a positive move away from the lack of acknowledgment that the disease has historically received. However, the coffers are not bottomless and with the front-loading of services there has been a relative absence of therapies and support to receive the influx of the newly diagnosed populations. Leading some to question the utility of early diagnosis of dementia (e.g., Fox et al., 2013) in the absence of sufficient post-diagnostic evidence-based therapies and demand more rigorous research.

The focus of research into Alzheimer's disease has historically predominated on profiling the cognitive impairments with significant investment into pharmacological options and the subsequent development of interventions such as cognitive rehabilitation (Clare & Woods, 2004) and cognitive stimulation therapy (Spector et al., 2003). While assessment and treatment of the cognitive impact of the dementias

---

* Originally published in 2014 in *Neuro-Disability & Psychotherapy*, 2(1/2): 80–99.

clearly has its role, the lived experience and meaning of being diag-nosed with dementia, that has repeatedly been spoken about in the qualitative literature (e.g., Clare, 2011), is often not addressed in a systematic way by service providers. The tides are turning, however, and the agenda is changing from chasing the diagnosis to post-diag-nosis. For example, as well as increasing their research budget to £5.2 million this year the Alzheimer's Society have recently split their fund-ing of research into two streams; with one being dedicated to "care, services and public health research". In the same vein the Faculty of the Psychology of Older People (FPOP) in conjunction with the Dementia Action Alliance have successfully requested of the Royal College of Psychiatry (February 2014) that post-diagnostic support become an obligatory rather than an optional standard within the Memory Service National Accreditation Programme (MSNAP).

Recent research has confirmed clinicians' observations that the onset and progression of dementia may pose a threat to a person's sense of identity (Clare, 2011). Following diagnosis many people describe being in a state of flux, experiencing both continuity and change in their sense of identity simultaneously (Clare, 2011). However, for some people the experience of a challenge to their pre-morbid identity is experienced as shame-inducing (Frank et al, 2006) and is met with self-criticism and social withdrawal. In a recent YouGov poll (cited in Cane & Cook, 2013), when asked if they lost friends after their diagnosis of dementia 40% of respondents said yes: for 12% "most of them", for 28% "some of them". At an individual level social withdrawal may be particularly evident where there is an emergence or re-emergence of "inadequacies" that were previously soothed through increased activity and observable achievements.

The magnitude of perceived stigma in people with Alzheimer's disease is comparable to or greater than other populations of people with chronic illnesses including cancer and Parkinson's disease (Burgener & Berger, 2008). This social amputation is recognised by the general public; with 59% of 2,070 members of the general public rating the inclusion of people with dementia in their community as bad (YouGov, 2011, cited in Cane & Cook, 2013). There are positive signs of reaction to the stigma and social exclusion. The Health Secretary, Jeremy Hunt, has praised those new businesses who have committed to training their staff in dementia awareness, raising the total number of volunteers to 250,000, and starting to give credence to the term "dementia friendly communities". The Health Secretary has stated that as a society "How we respond to dementia is the litmus test of whether we can face up to the challenge of an ageing population, and do so in a way which allows compassion and dignity" (Hunt cited by Donnelly, 2014).

Compassion is a theme that is common to many interpersonal and intrapersonal therapies and a cornerstone of many religions. It is the aim of services and the hope of individual care professionals that our services are delivered in a compassionate way, but compassion focused therapy (CFT) is a specific model of therapy that can structure a systematic delivery of compassion-based services. Compassion-based care has been much talked about within the context of NHS services in the wake of the Francis report (The Mid Staffordshire NHS Foundation Trust Public Inquiry, 2013). The definition of compassion within the CFT model is "a basic kindness, with a deep awareness of the suffering of oneself and of other living things, coupled with the wish and effort to relieve it" (Gilbert, 2009, p. xiii).

CFT "is rooted in an evolutionary, neuro- and psychological science model" (Gilbert, 2010, p. 9), and the conceptualisation of compassion is also rooted in evolutionary terms; focusing on the capacity of humans to form attachment bonds, engage in kinship caring and social co-operation. Humans are born altricial; being incapable of moving around independently at birth and requiring nourishment to be provided in order for them to survive. This dependency demands a bond be created with an independent other; typically the mother. When a safe bond is created this results in a soothing experience for the human infant. Conversely being without a bond is unsafe in survival terms and therefore results in a distressing experience for the human infant that drives him/her to re-attach to his/her "parent". Within the delivery of CFT there is an element of psycho-education, part of which is to explain these concepts; that as humans we have in-built in us essential capacities to experience distress, but also capacities to detect feelings of warmth and safety and be soothed and to soothe others.

Gilbert's model (2005, 2010) builds on these in-built drivers to create CFT, drawing on a number of treatment models, such as cognitive behavioural therapy (CBT) and Linehan's dialectical behaviour therapy (DBT) (Read, 2013). CFT is a therapeutic model where the key components are care for the well-being of others, sensitivity to distress, sympathy, distress tolerance, empathy, and a non-judgemental stance towards experiences, which was developed for people with chronic and complex mental health problems linked to shame and self-criticism (Gilbert, 2010). In particular it developed as a response to those patients who were struggling with the predominating therapy (CBT), many of whom stated that they could understand the model at a rational level, but could not "feel" appropriate changes. CFT explains why this phenomenon should exist; that the human brain has evolved from a more primitive level of functioning that continues to exert motives and regulate our emotions, in contrast the more newly evolved part of the brain hosts the reasoning, thinking, and self-reflection. Co-ordinating these two is "tricky", but can be made more tolerable if the evolutionary functions are brought to our awareness, accepted and understood as not being our fault, but allowing ourselves to live with more compassion and free of shame is our responsibility. This gives rise to the concept that "it's not your fault, but it is your responsibility", which is shared with the patient throughout the therapy, to remove blame from the patient for their emotional experiences but to encourage them and promote their sense of agency in their future.

The pro-social requirement of being human provides us with the capacity for understanding the intentions of others. Depending on how our brains are regulating our emotions at a given moment will determine whether we are more likely to perceive threatening intentions, competitive intentions, or compassionate intentions from others. Although this can function to moderate our behaviour to improve our relationships it also provides the capacity to perceive the self in the other's mind. This can give rise to the concept of shame. CFT distinguishes the roles of internal and external shame. Internal shame is where the individual negatively focuses on their own "inadequacies" in their own mind, for example, "I am useless". External shame is related to how the individual believes he/she is perceived in the mind of the other person, for example, "I think they think I am useless".

Research into the neurophysiology of emotion suggests that we can distinguish at least three types of emotion regulation system (Depue & Morrone-Strupinsky, 2005): threat and protection systems; drive, resource-seeking, and excitement systems; and contentment, soothing, and safeness systems. CFT focuses on the interplay between threat, motivational, and soothing psychobiological systems. These systems have been simplified into the three circles diagram; which, in the author's clinical experience, most patients even with mild dementia find understandable (see Figure 1).

The threat system is acknowledged as being the most dominant; if there is any sense of doubt the threat system will "trump" the others, because at an evolutionary level the most important factor is survival. The threat system can, however, be regulated by the other positive systems. One way of down-regulating the threat system is through the activation system, for example by working, going to the gym, etc., a coping strategy that may be very effective through adult life. The other system available to down-regulate the threat system is the calming and soothing system, which can be engaged, for example, by being hugged, practicing meditation, or mindful breathing.

As noted previously CFT draws on other well developed models, including CBT and DBT, and parallels can be drawn with the formulations of these models. The formulation within CFT goes through a series of stages (Gilbert, 2012):

Background and historical influences: exploring basic early attachment styles and life events, as well as key emotional memories. This process helps to develop a coherent narrative within a containing, non-judgmental setting.

Figure 1: Types of affect regulator systems.

Key threats and fears: Within CFT, there is a distinction made between external and internal threats; where external threats refer to what the world or others might do, and internal threats are related to what emerges or is recreated inside oneself.

Safety strategies: As the threats can be differentiated into internal and external, so the safety strategies can be targeted at the internal world or the external world. Some strategies can be automatic, linked to classical conditioning, where others are more planned.

Unintended consequences: The strategies aimed at keeping the individual safe from their key threats can often lead to unintended, unhelpful consequences that can prevent them from realising their own values and life goals.

The formulation focuses on de-shaming through the education that the evolutionary design of the brain makes it function in ways that is not are not the fault of the individual, repeating again the message that "it's not your fault, but it is your responsibility".

The application of CFT is widening with reports of good effect in populations including those experiencing trauma (Lee, 2012; Troop & Hiskey, 2013), psychosis (Braehler et al., 2012), acquired brain injury (Ashworth et al., 2011), eating disorders (Goss & Allan, 2014), and for groups of patients with mixed diagnoses residing on inpatient wards (Heriot-Maitland et al., 2014). There have also been preliminary reports of the efficacy of CBT with a compassionate focus when used with people with mild dementia and depression (Green, 2007).

The relevance of CFT to dementia: dementia as a threat

Dementia represents a very real threat; "the dreaded 'A' word" as one older lady told me of her fear of Alzheimer's disease. It is a threat to life, a threat to function, a threat to finances, a threat to self-identity, and a threat to social inclusion. In addition it will become a threat to the activating system described above, reducing the ability of and ultimately preventing people with dementia from performing activities such as driving, working and physical exercise. Dementia therefore reduces the opportunity to use activation as a coping strategy to down-regulate the threat system. The very experience of being diagnosed with dementia can be perceived as a threat. Both the diagnostic process when relatives are asked to give their perspectives on the person's memory can be threat inducing: "I feel as though everyone is ganging up on me . . . There's two against one" (Mr B, 2013), as much as receiving the diagnosis itself: "It comes down on you like a ton of bricks" (Mrs B, 2014).

One of the central evolutionary tenets to CFT is that the human is born a pro-social creature with a need to form immediate attachments in order to survive, and with an ability to interpret social situations. The presence of an attachment figure results in feelings of safeness, whereas events that rupture these feelings of safeness are perceived as threatening. The body of research defining the role of attachment in dementia is significant; for example, the work of Meisen theorises that dementia is a loss process that activates the experience of feeling unsafe and the emotional need for the security of an attachment figure (Osborn et al., 2010). Hence the experience of unsafeness triggered by dementia can activate an over-reliance on threat-based strategies to regulate one's feelings, resulting in what may be labelled as "challenging behaviour".

As we are social animals, in order to survive we have to function within social packs. The central goal is survival of the gene, rejection from the social pack threatens our genetic survival and so we have developed systems to evaluate social processes and avoid rejection regardless of our biological age.

When members of the public are asked for their immediate images of dementia, they tap into the cognitive impairments such as "fog" and "confusion", the social impairments such as "loneliness", and sometimes they can tap other senses such as "smelly". On receiving any diagnosis, many people these days will return home to search for information on the internet. The images that are generated by internet search engines are no more favourable than those described above reflecting visual images of despair. The jump from the public perception of dementia to J. K. Rowling's use of the characters of dementors is not so great.

> Terrible things they are (p. 70) . . . glistening greyish and scabbed, like something dead that had decayed in water . . . (p. 66). They suck the happiness out of a place, Dementors (p. 76) (Rowling, 1999)

Rationale for increasing compassion in the person with dementia

> It's not pretty where you're going
>
> Mr J, 2014

Shame and its antidote compassion have now been widely evidenced, across a range of mental health diagnoses, as noted previously, and brain injury, to be a robust construct predictive of outcome and malleable to CFT intervention. The CFT model is a trans-diagnostic approach that uses the formulation as a starting point and works with the process that the individual is experiencing, rather than starting with a "diagnosis". It enables a shared formulation to be held where the experienced pain is validated. It can give explanation for those cases where the extent of the pain may not be mirrored by a significant decline in measured cognitive function. It enables patients to understand and make sense of core beliefs and to learn to identify shame-based beliefs. It distinguishes responsibility from blame and offers the option of a positive way of improving the lived experience, using techniques based on developing and nurturing a sense of compassion to challenge the shame often experienced.

Adherence to the recovery model and Implementing Organisational Change through Recovery is a current pledge of many NHS services. Consistent with the application of the recovery model within dementia care (Hill et al., 2010) the emphasis of CFT is on the process rather than the outcome. Recovery within dementia can be conceptualised as "a process, a way of life, an attitude and a way of approaching the day's challenges" Deegan (1996).

Advocates of the person-centred approach argue that emotions and interpersonal responses may remain intact for significantly longer into the disease process than the cognitive skills. A recent study has provided neuropsychological and neuro-anatomical support for the personcentred approach; demonstrating, for the first time, preserved complex emotion-based learning capacity, despite profound episodic memory impairment in Alzheimer's disease (Evans-Roberts & Turnbull, 2011). These findings are consistent with the lived experience described, for example, by Keith Oliver, Alzheimer's Society envoy. "I cannot hang on to what people say to me but I can hang on to how they make me feel" (Oliver, 2014).

Therefore, the possibility for eliciting change based on emotional learning appears viable even with people with more advanced dementia. At an anatomical level, compassion has been associated with activation of the periaqueductal grey matter (PAG) visible on functional magnetic resonance imaging (fMRI) (Emiliana et al., 2012). The PAG lies within the midbrain and is also associated with pain, perception of pain in others, and maternal behaviour. As part of the psycho-education aspect of CFT, patients with dementia can find it very helpful to hear, in simplistic terms, that the PAG is a different part of the brain to that classically associated with degeneration in the early stages of, for example, Alzheimer's disease, encouraging hope and progress in a narrative that is often dominated by loss, degeneration, and disability. The CFT model also encourages the acquisition of a concrete and observable skill: that of soothing rhythm breathing. Again the explicit conversation that the therapist holds an expectation that the person with dementia will be learning and acquiring new skills in the process of the CFT is an antithesis to the societally held narrative of hopeless decline and degeneration.

So far we have considered compassion as an antidote to "self-to-self" shame. However, within the CFT framework the flow of compassion (and equally the flow of shame) is simultaneously considered in terms of "self-to-other" and "other-to-self".

Rationale for increasing compassion in the care-giver

He forgets everything . . . he has a distorted view of life

comments made during assessment by Ms R sister of Mr R who was diagnosed only with mild cognitive impairment (MCI).

There's no point me asking him . . . he's no use anymore

comments made by Mrs G wife of Mr G in the early stages post-diagnosis. Mr G was at this stage still working in a job of very high societal standing.

"Other-to-self" shame, which can also be thought about as received criticism, can be operationalised as expressed emotion (EE). EE is a construct that has been heavily researched across a broad range of psychiatric conditions where it has been shown to be a robust predictor of illness prognosis (Wearden et al., 2000) and neurodegenerative disorders, and encompasses emotional over-involvement, criticism, and hostility. The concept of EE can be unhelpfully construed in a unidirectional way that can be perceived as blaming of the relative, however, it can more helpfully be construed as a symptom of the relational milieu being experienced by both people in the relationship dyad, and recent research has elucidated the links between high EE and higher levels of both shame and guilt/self-blame in the relative (Wasserman et al., 2012). Higher levels of EE have been associated with higher levels of care-giver distress in the carers of people with dementia (Tarrier et al., 2002). EE has also been associated with behavioural symptoms in the dementia care recipient and a number of non-cognitive features including physical aggression, anger, threatening, uncooperative and paranoid behaviour, and wandering (Tarrier et al., 2002). These findings remind us that some symptoms of dementia are affected by the care environment and not solely by the biological disease process, as highlighted by Kitwood two decades ago (Kitwood, 1997).

Cooney and colleagues (2006) identified that a greater level of EE in carers of people with dementia is highly correlated with all forms of abuse (verbal, physical, and neglect) and recommended prioritising management of high EE and psychological distress in carers, treating behavioural problems in the dementia sufferer, and educating carers in managing behavioural problems in their dependants. In the general population, 6% of older people have experienced abuse in the last month and this rises to approximately 25% in vulnerable populations such as people with dementia (Selwood & Cooper, 2009).

Over a third of family carers reported significant abuse from the people they cared for. Carers who reported more abuse also reported a greater deterioration in their relationship with the person with dementia, and were themselves more anxious and depressed. The extent to which carers used dysfunctional coping strategies partially explained this, suggesting that interventions to change the carers' coping styles might alleviate the impact of abusive behaviour (Cooper et al, 2010). Although EE does not seem to affect the rate of cognitive decline, the behavioural symptoms of dementia have been described as being more distressing to carers than the cognitive symptoms, and to dictate the decision to institutionalise the care-recipient (Vitaliano et al., 1993).

Offering CFT in this example can be used not only to aid the caregiver in understanding the dementia, but also to empathise with the person with dementia. Specific techniques can be taught to help them to manage the stress associated with living with the person with dementia, and to be self-compassionate towards themselves when they experience feelings of hostility towards their loved ones, normalising this experience. Within the CFT model it is understood that factors that identify difference are more associated with social rank differentiation, dominance and subjugation, whereas factors that identify commonality are more associated with affiliation and compassion. Using the CFT model post-diagnosis can enable both the carer and person with dementia to join together in the experience of feeling bewildered with calmness and compassion for themselves and each other, a response that remains appropriate as the disease progresses. This is the antithesis to the usual post-diagnostic response in which difference is immediately imposed on the couple's relationship with one person being labelled as a patient and the other being labelled as a carer, with the assumption that the difference between the two will progressively increase.

Case study

Mr W was a fifty-eight-year-old white British man, pre-morbidly academically gifted and estimated to have functioned optimally in the "superior" range, who had been referred to the early onset dementia clinic, with cognitive decline He was labelled at referral as being a "complex case" that included a history of significant alcohol abuse, drinking up to two bottles of vodka a day at the height of his drinking, and a previous history of depression. Mr W also presented with complex interpersonal relationships including "uncontrollable" anger and an explosive temper that was considered to put others at danger. Mr W was considered to have limited insight into both his interpersonal and cognitive limitations. He had been previously treated in the private sector with CBT, but was reported not to have derived benefit from this, and described the experience as "everything happens too quickly".

The author was initially involved in the formal neuropsychological assessment of Mr W. This revealed a scattered profile across cognitive domains, with processing speed, executive function, and memory being most affected. In comparison to the general population norms Mr W was functioning from the ninth to the ninety-fifth percentile; across the superior, high average, average, low average, and borderline ranges. However, as he was estimated to have functioned in the superior range pre-morbidly, these represented personal declines of up to three standard deviations. One of the challenges of this case was a lack of agreement within the service on Mr W's diagnosis resulting in a perpetuation of uncertainty for the family and an experience of being a "political hot-potato".

The development of the formulation identified that Mr W had experienced his childhood as positive, with an absence of any disclosed abuse. He described his mother as having a good sense of humour, being compassionate to others, and kind and generous in her actions. He was the third child in a sibship of six boys, his father worked very long hours and as such parental attention was limited. Within the culture he grew up in physical displays of emotion were typical, and the ability to defend oneself physically was important. Mr W became adept at utilising his drive system to avoid his threat system, and to boost his self-esteem, and ultimately held a national role in a large organisation. However, with the onset of cognitive challenges, as he found his job and social roles more challenging, he was less able to access his resource-focused drive system, and found less helpful ways of regulating his threat system including verbal aggression, fighting, and fleeing via alcohol.

The emotional issues that arose for Mr W as a result of his cognitive difficulties resulted in continual negative self-commentary such as:

Setting myself too many hurdles . . . I've failed. I feel belittled.

These examples show not only that Mr W's cognitive limitations resulted in self-to-self criticism (internal shame), but also in the belief of other-to-self criticism (external shame).

CFT was introduced to Mr W and simplified using the model of the three-circles. Mr W was supported to make sense of his experience in terms of the CFT model. He was able to understand the CFT concepts of the old brain and the new brain and how our current behaviours can be understood with evolutionary functional analysis. Mr W was encouraged to notice that his current level of cognitive functioning was, in normative terms, good, however, it represented a personal decline for him, and it was his response to that personal decline that was disabling him.

At the initiation of therapy Mr W's coping strategies were to be verbally abusive to both his wife and to strangers, and to be physically aggressive within the home. The three circles model was used to identify that his cognitive decline was being perceived by him as a threat, specifically a threat to his social acceptance, his social status, and his social inclusion, as well as being a threat to his ability to provide for and protect his family. Mr W was supported to understand that his current attempts at coping, that is, the physical and verbal expression of anger, were triggered threat responses (see Figure 2).

Mr W was able to understand that a response triggered from the "compassion circle" could be an alternative, but he commented that he did not think he currently had one. However, he was keen to try to develop his compassionate self and chose to personalise the three circles model. Typically the circles are presented as coloured, with the threat-focused system being represented in red, the resource-focused system being represented in blue, and the affiliate-focused system being represented in green. Mr W however felt that his affiliate-focused system would be best represented as "pink and fluffy and warm", and this was conveyed with sincerity, and so was adopted for the remainder of the therapy.

Mr W had described his mother as a person who was "very kind and caring" and, reflecting on the role of the PAG reported above, possibly as a result of having experienced maternal nurturance in his childhood, he found it relatively easy to generate compassionate images for use within the CFT. He was also able to learn to use the soothing rhythm breathing technique with as much ease as patients without dementia that I have worked with.

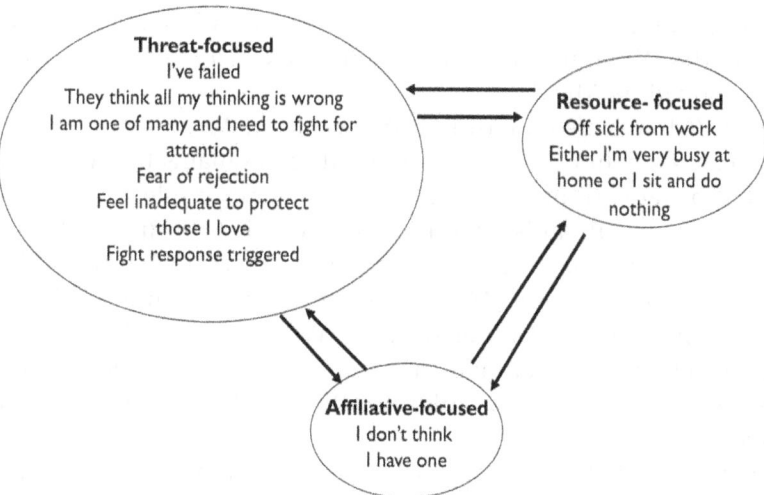

Figure 2: Diagrammatic formulation of Mr W's affect regulator systems.

It was identified by the neuropsychological assessment that Mr W was experiencing greatest personal decline in the domain of processing speed, and he had made reference to his experience of being outpaced by the, presumably non-modified, CBT that he had engaged in. It was important, therefore, to deliver the CFT in a way that was adapted to Mr W's processing speed by slowing the delivery of the concepts and repeatedly checking back in with Mr W, in a non-shaming way, that the therapy was making sense to him. Adapting CFT in this way sits comfortably conceptually with the model itself, as Paul Gilbert has noted "When training people from other approaches, particularly CBT, we find that we have to slow them down; to allow spaces and silences for reflection and experiencing within the therapy . . ." (Gilbert, 2010, p. 7). However, it could be noted here that for other patients with dementia that I have worked with, whose decline was more significant in the domain of memory, I have adapted the therapy to reduce the length of silences to reduce the experienced threat of being abandoned in the silence.

The CFT was also adapted to account for Mr W's memory decline, this included using a white board throughout each therapy session to provide an evolving visual record of the session, providing a written copy at the end of each session of the whiteboard material (other patients I have worked with have used the camera function on their smartphone to record their whiteboard work), the provision of handouts regarding specific cognitive strategies, and ultimately the inclusion of Mrs W within the sessions could be considered to reduce the memory load on Mr W.

The CFT was also mindful of the results of Mr W's neuropsychological assessment by aiming to utilise those domains that were identified as being current strengths for Mr W. For example, his identified strength in visual skills gave rise to an explicit discussion that the therapist considered the use of visual imagery to be particularly accessible and useful to Mr W. The development and application of visual imagery was adopted by Mr W as a self-soothing strategy to good effect, for example, he created images of people in the street outside smiling and children sitting in the nearby burger cafe laughing.

Although Mr W was able to understand the model and found it easy conceptually, he struggled with the process; some of his struggles were similar to those that one would expect within any intrapersonal therapy requiring reflection and change. For example, "You can't think of all these three when you're in it. You're just stuck in one," which reflects the struggle of being trapped within a process rather than being able to reflect on the trap and the alternative contemporaneously; the development of meta-cognition. Other struggles were specific to the ethos of this model of CFT. For example, he spoke of his concern that if he were to respond to what he perceived to be an interpersonal challenge from the affiliate-focus system, "It's like rolling over and saying you can do it again."

However, this type of concern is a commonly raised challenge within the early stages of CFT and is not specific to its use within a dementia population (Gilbert & Irons, 2012, personal communication).

As Mr W's insight increased throughout the therapy, it resulted in an increased awareness of the role of responsibility. In Mr W's particular case he became aware of the role that smoking and alcohol had in his interpersonal relationships, but also in his cognitive functioning. He spoke of "kicking myself now for what's happened". However, the model of CFT effectively manages this issue and differentiates the blame for past events from responsibility for present and future self; one of the key messages of CFT being "it's not your fault but it is your responsibility". And Mr W's self-blame softened somewhat, commenting "You can beat yourself up but it won't change the past."

At the points when Mr W raised his fears of his involvement in causing his cognitive problems, we explicitly referred back to the CFT model and Mr W was encouraged to reflect on this; to identify which of his response systems was being triggered and to identify how he could achieve more balance between the three emotion regulation systems.

Mr W had spoken about his concerns of being in public and saying the wrong thing, or of being put on the spot, which would highlight his "inadequacies". Through the therapy Mr W was able to come to the understanding that, "Anger is the result of my feelings—it's not about other people. I can't change what they say only how I interpret it." He had spoken of his fears of not being able to converse properly, "I can't get it out of my mouth . . . I can think it through in my head . . . I walk away and I can't get the words out. I feel belittled."

By considering that conversation may serve a social role, which could serve an affiliative function, Mr W was encouraged to wonder how it would be possible to remain with his affiliate-focused regulation system being dominant. He identified how he would act if the roles were reversed, that he would treat the person with dementia normally, slow down, bear with it, and he also identified that the problem would be more tolerable if he allowed himself to share the burden of communication with the listener. As a result he felt more confident both at going out in public again, and also of telling his closer friends of his dementia, which resulted in positive experiences for him.

The therapy progressed and Mr W became more confident with the CFT model, and he chose to involve his wife in the fifth therapy session. This represented a big shift for him from trying to hide his limitations and experiencing external gaze as shaming—literally fighting those he felt gazed shamefully upon him—to a point where he felt safe enough to invite external opinion and speak of his difficulties. Due to the simplicity of the three circles model it was possible to introduce it to Mrs W within this joint session, which she understood, and it was used as an externalising tool. One of the concerns they raised was how to cope when Mr W repeatedly asked questions. The couple were asked to reflect on their thoughts when this happened. Mr W stated, "I think she's hiding information from me". Mrs W stated, "I feel as though he's accusing me of keeping things from him".

And then they would spiral into a "Yes you are, No I'm not" argument. The couple were asked instead to consider their dilemma from the viewpoint of the three circles model and consider "when a question is repeatedly asked where are you both on the model?". They were able to identify that they were both functioning from within their threat system, which gave them the sense of "we're in this together", and they identified that an exit strategy could be via the compassion system. At the ending of therapy the couple reported, "We're having a lot more fun . . . holding hands."

Mrs W also reported that she could "visibly see when (Mr W) is checking himself" to reduce his aggressive outbursts, and the frequency of their arguments reduced.

From Mr W's perspective by the end of therapy he reported "I'm less in here now" (pointing to the threat-focused circle on the model) and he felt that he was also more able to access his motivational system, reporting that he was "able to see pleasure in lots of things, in small things" such as playing with their dogs, and working out on their small-holding. By the end of therapy there was evidence of a reduction in both self-to-self and self-to-other criticism. He had stopped avoiding speaking on the phone, which prior to therapy was triggered by the external shame that "they think all your thinking is wrong", in turn triggering an anger response of "who are they to judge me". Instead he tried to soothe himself, by what he described as "tuning in to the preparation" through a number of steps including soothing rhythm breathing, making a joke of himself, taking his time, and paying attention. Mr W's ability to access his affiliate-focused system more readily resulted in a reduction in his experience of shame, and hence he wanted to socialise more, to go out to the pictures, and go out for drinks with neighbours. He continued to experience difficulties if more than one person spoke at a time, and with switching his attention between people in conversations; however, he was able to respond to his experience in a more helpful way, rather than retreating into fear and social isolation.

Discussion

CFT is considered to be a "third wave" approach to psychological therapy, and is becoming increasingly widely applied to a range of mental health conditions and neurological disorders. This chapter suggests that CFT can be applied also to populations with neurode-generative conditions such as the dementias. The model of CFT with its focus on evolutionary functional analysis responds particularly well to the existential threat intrinsic in a diagnosis of dementia. The concepts can be simplified sufficiently well to be understood by patients with mild dementia, and yet still remain clinically meaning-ful. The three circles model, which can be presented in visual, verbal, and/or physical modalities, accommodates for limitations within a variety of cognitive domains, and appears to be sufficiently tangible and concrete to allow for the less flexible and less abstract thinking of dementia. Mr W, whose case was presented here, when functioning to his optimum pre-morbidly was estimated to have functioned in the "superior" range. Arguably he entered the dementia process with significant cognitive reserve, and although his skills in some areas had dropped by three standard deviations, this still left him functioning in the "low average" range in comparison to the general population. It could therefore be questioned as to how applicable CFT really is to people who have less cognitive reserve and are more advanced in their dementia. In the author's clinical experience as the severity of dementia progresses the conceptual accessibility of the CFT model declines, which one would expect with the acquisition of any conceptual information, however, the more concrete tasks central to the compassionate mind training (CMT), such as the skill of breathing mindfully, appear to be possible to acquire into the moderate stages of dementia. Most recently two men (with Addenbrooke's Cognitive Examination-III scores of sixty-eight and fifty-five respectively), both experiencing moderately severe dementia, took part in a six-week group format of CFT, with their spouses, and were able to acquire the soothing rhythm breathing skill, with a resultant reduction in their breathing rates from session one to session six. The need for CMT skills to be practised repetitively across multiple occasions lends itself well to people's learning abilities in the early stages of dementia.

It could almost be said that CFT has "it's not my fault, but it's my responsibility" as its strap-line, which can be particularly helpful in response to some people who may feel greater shame as a result of considering that their lifestyle was to blame for their dementia.

Working therapeutically involves not just the experience of the patient, but also the experience of the therapist. CFT was chosen to be used with Mr W partly due to his previous poor response to CBT. However, working within the model of CFT was experienced as particularly containing by the author with this case, which raised concern within the providing system, triggering anxiety among clinicians when a differential diagnosis could not be agreed upon. The model helped the author to be mindful of the seductive draw towards the "threat zone", to reflect in an evolutionary functional analysis way on the meaning and experience of being unable to arrive at a definitive diagnosis within a social hierarchy and the impact this has on fear, such as fear of exposure. The CFT model then enabled the author to respond to these threats with self-to-self compassionate responses, and hopefully with self-to-other compassionate responses when working with colleagues on Mr W's case.

The model of CFT facilitates the therapist and the patient to come together in acknowledging and sharing of core human experiences: fear, anxiety, anger. For example, the author was mindful of the felt experience of being referred a large, physically fit, and relatively young man who was described as explosive and violent, for therapy in a one-to-one setting. This experience was made more tolerable by acknowledging that Mr W was simultaneously experiencing social threat and fear. Acknowledging and working with shared emotional states within CFT is possible in the present, in contrast to some therapies where the reliance on having to reflect on poorly remembered experiences or thought processes that would differentiate the patient from the therapist.

Although the body of research is increasing in the applications of CFT, it has been cautioned that as an intervention CFT has not yet been fully subjected to extensive, rigorous evaluation (Beaumont et al., 2012). As with any intrapersonal therapy an ability to engage with the therapy requires some shared aims and a degree of insight. Insight is an interesting concept, with anosognosia often being noted as one of the sequelae of both acquired brain injury and neurodegenerative disorders, but lack of insight can present in an almost identical way to denial, and hence the psychological formulation is crucial. CFT uses evolutionary functional analysis to produce its formulation, if denial is serving a self-protective function from threat, the person would need to perceive of an equally self-protective alternative for insight to be allowed to emerge as a non-shaming experience. Within the model of CFT the alternative would be to regulate these threat-focused emotions by experiencing emotions from the soothing system. It is therefore important that insight is not solely considered to be a direct cause of cognitive limitations, which can result in referrers excluding people with cognitive challenges from the opportunity to receive therapy, indeed in this case Mr W's insight improved as a function of having engaged in the CFT. However, in reality there is a reduction in insight with the progression of dementia in its later stages, and it is probable that the way in which CFT has been applied within this case study would be less appropriate for people with a dementia significantly limiting their insight, however, some of the more procedural skills, such as mindful breathing, may be attainable. This deserves further attention.

The man in this case study chose to bring his wife into the fifth session with a positive effect for himself, his wife, and their relationship. CFT appears to be able to function well for couples living with a diagnosis of dementia and we are currently piloting providing CFT to couples post-diagnosis in a group format.

References

Ashworth, F., Gracey, F., & Gilbert, P. (2011). Compassion focused therapy after traumatic brain injury: theoretical foundations and a case illustration. *Brain Impairment, 12*(2): 128–139.

Beaumont, E., Galpin, A., & Jenkins, P. (2012). Being kinder to myself: a prospective comparative study, exploring post-trauma therapy outcome measures, for two groups of clients, receiving either cognitive behaviour therapy or cognitive behaviour therapy and compassionate mind training. *Counselling Psychology Review, 27*(1): 31–43.

Braehler, C., Gumley, A., Harper, J., Wallace, S., Norrie, J., & Gilbert, P. (2012). Exploring the change processes in compassion focused therapy in psychosis: results of a feasibility randomized controlled trial. *British Journal of Clinical Psychology, 51*: 1–16.

Burgener, S., & Berger, S. (2008). Measuring perceived stigma in persons with progressive neurological disease: Alzheimer's dementia and Parkinson's disease. *Dementia: The International Journal of Social Research and Practice, 7*(1): 31–53.

Cane, M., & Cook, L. (2013). *Dementia 2013: The Hidden Voice of Loneliness.* London: Alzheimer's Society.

Clare, L. (2011). I'm still the same person. *Dementia, 10*(3): 379–398.

Clare, L., & Woods, R. T. (2004). Cognitive training and cognitive rehabilitation for people with early stage Alzheimer's disease: a review. *Neuropsychological Rehabilitation: An International Journal, 14*(4): 385–401.

Cooney, C., Howard, R., & Lawlor, B. (2006). Abuse of vulnerable people with dementia by their carers: can we identify those most at risk? *International Journal of Geriatric Psychiatry, 21*: 564–571.

Cooper, C., Selwood, A., Blanchard, M., Walker, Z., Blizard, R., & Livingston, G. (2010). The determinants of family carers' abusive behaviour to people with dementia: results of the CARD study. *Journal of Affective Disorders, 121*: 136–142.

Deegan, P. (1996). Recovery as a journey of the heart. *Psychiatric Rehabilitation Journal, 19*: 91–97.

Depue, R. A. & Morrone-Strupinsky, J. V. (2005). A neurobehavioural model of affilitative bonding. *Behavioural and Brain Sciences, 28*: 313–395.

Donnelly, L. (2014). Jeremy Hunt promises revolution in care for dementia sufferers. Available at: www.telegraph.co.uk/health/nhs/10666591/Jeremy-Hunt-promises-revolution-in-care-for-dementia-sufferers.html

Emiliana, S-T., Godzik, R., Castle, J., Antonenko, E., Ponz, O., Kogan, A., Keltner, A., & Dacher, J. (2012). An fMRI study of caring vs self-focus during induced compassion and pride. *Social Cognitive and Affective Neuroscience, 7*(6): 635–648.

Evans-Roberts, C., & Turnbull, O. (2011). Remembering relationships: preserved in motion-based learning in Alzheimer's disease. *Experimental Aging Research, 37*(1): 1–16.

Fox, C., Lafortune, L., Boustani, M., & Brayne, C. (2013). The pros and cons of early diagnosis in dementia. *British Journal of General Practice, 63*(612): e510–e512.

Frank, L., Lloyd, A., Flynn, J., Kleinman, L., Matza, L., Margolis, M., Bowman, L., & Bullock, R. (2006). Impact of cognitive impairment on mild dementia patients and mild cognitive impairment patients and their informants. *International Psychogeriatrics, 18*(1): 151–162.

Gilbert, P. (2005). Compassion and cruelty: a biopsychosocial approach. In: P. Gilbert (Ed.), *Compassion: Conceptualisations, Research and Use in Psychotherapy* (pp. 9–74). London: Routledge.

Gilbert, P. (2009). *The Compassionate Mind*. London: Constable & Robinson.

Gilbert, P. (2010). *Compassion Focused Therapy: Distinctive Features*. London: Routledge.

Gilbert, P. (2012). *Compassion Focused Therapy: Three Day Advanced Workshop with Prof Paul Gilbert*. Derby: Compassionate Mind Foundation.

Goss, K., & Allan, S. (2014). The development and application of compassion-focussed therapy for eating disorders (CFT-E). *British Journal of Clinical Psychology, 53*: 62–77.

Green, P. (2007). Adapting CBT using a compassionate mind approach with older people who have dementia and depression. *PSIGE Newsletter, 99*: 5–8.

Heriot-Maitland, C., Vidal, J. B., Ball, S., & Irons, C. (2014). A compassionate-focused therapy group approach for acute inpatients: feasibility, initial pilot outcome data and recommendations. *British Journal of Clinical Psychology, 53*: 78–94.

Hill, L., Roberts, G., Wildgoose, J., Perkins, R., & Hahn, S. (2010). Recovery and person-centred care in dementia: common purpose, common practice? *Advances in Psychiatric Treatment, 16*: 288–298.

Kitwood, T. (1997). *Dementia Reconsidered: The Person Comes First*. Buckingham: Open University Press.

Lee, D. (2012). *The Compassionate Mind Approach to Recovering from Trauma using Compassion Focused Therapy*. London: Robinson.

Mid Staffordshire NHS Foundation Trust Public Inquiry (2013). *Report of the Mid Staffordshire NHS Foundation Trust Public Inquiry Chaired by Robert Francis QC*. Available at www.midstaffspublicenquiry.com

Oliver, K. (2014). Using psychological support. Paper presented at *What's Happening Now? Psychological and Psychosocial Aspects of Dementia and Dementia Care.* The British Psychological Society Faculty of the Psychology of Older People Training Day, 15 January 2014.

Osborn, H., Stokes, G., & Simpson, J. (2010). A psychosocial model of parent fixation in people with dementia: the role of personality and attachment. *Aging & Mental Health, 14*(8): 928–937.

Read, K. (2013). *Instructor's Manual for Dialectical Behaviour Therapy with Marsha Linehan.* Mill Valley, CA: Psychotherapy.net.

Rowling, J. K. (1999). *Harry Potter and the Prisoner of Azkaban.* London: Bloomsbury.

Selwood, A., & Cooper, C. (2009). Abuse of people with dementia. *Reviews in Clinical Gerontology, 19*(1): 35–43.

Spector, A., Thorgrimsen, L., Woods, B., Royan, L., Davies, S., Butterworth, M., & Orrell, M. (2003). Efficacy of an evidence-based cognitive stimulation therapy programme for people with dementia: randomised control trial. *British Journal of Psychiatry, 183*: 248–254.

Tarrier, N., Barrowclough, C., Ward, J., Donaldson, C., Burns, A., & Gregg, L. (2002). Expressed emotion and attributions in the carers of patients with Alzheimer's disease: the effect on carer burden. *Journal of Abnormal Psychology, 111*(2): 340–349.

Troop, N. A., & Hiskey, S. (2013). Social defeat and PTSD symptoms following trauma. *British Journal of Clinical Psychology, 52*: 365–379.

Vitaliano, P. P., Young, H. M., Russo, J., Romano, J., & Magana-Amato, A. (1993). Does expressed emotion in spouses predict subsequent problems among care recipients with Alzheimer's disease? *Journal of Gerontology, 48*(4): 202–209.

Wasserman, S., Weisman de Mamani, A., & Suro, G. (2012). Shame and guilt/self-blame as predictors of expressed emotion in family members of patients with schizophrenia. *Psychiatry Research, 196*(1): 27–31.

Wearden, A. J., Tarrier, N., Barrowclough, C., Zastowny, T. R., & Rahill, A. A. (2000). A review of expressed emotion research in health care. *Clinical Psychology Review, 20*(5): 633–666.

CHAPTER SIX

The use of yoga to enhance a compassion focused therapy intervention in a holistic neuropsychological rehabilitation setting for ABI: a qualitative case illustration*

Aneesh Shravat

Introduction

Why yoga in a holistic neuropsychological rehabilitation centre?

Yoga is an ancient practice that originated in India and involves controlled breathing and relaxation techniques as well as undertaking stretches and both standing and seated poses. Studies have comprehensively demonstrated the benefits of yoga in order to help alleviate mood difficulties. For example, Michalsen and colleagues (2005), highlight that over a three month period weekly classes of yoga can have a significant impact on reducing feelings of stress and depression. Vancampfort and colleagues (2011), also highlight that Hatha yoga sessions that focus on the present and relaxation techniques can lead to significant reductions in anxiety and psychological stress. Smith and colleagues (2007), state that the use of yoga can help offer an alternative modality for rehabilitation in terms of helping reduce symptoms of stress and anxiety by focusing on breathing awareness and internal concentration to remove external concerns in patients in a health setting. Indeed, Lavey and colleagues (2005), also found that regular yoga

---

* Originally published in 2014 in *Neuro-Disability & Psychotherapy*, 2(1/2): 100–107.

helped to alleviate difficulties with mood for patients in a psychiatric inpatient unit. The use of yoga in stroke rehabilitation can be beneficial for depression and increase a sense of well-being for people who have had a stroke (Bastille & Gill-Body, 2004; Lynton et al., 2007). In addition, Lundgren and colleagues (2008) undertook a randomised controlled trial focusing on the use of acceptance and commitment therapy (ACT) and yoga for drug-refractory epilepsy. The participants that attended the yoga sessions as well as ACT were found to have decreased seizure index scores and increased quality of life.

Compassion focused therapy (CFT)

The emotional consequences of brain injury have been well documented with neurological, psychological, and social aspects increasingly well researched and defined (Wilson & Gracey, 2009). The compassion focused therapy (CFT) approach places emphasis on the emotional experience associated with psychological problems. This model draws on social, evolutionary (especially attachment theory), and neurophysiological approaches to affect regulation (Gilbert, 2010). The model is based upon the fact that attachment and affiliative behaviours have evolved over many millions of years to regulate threat based emotions and action tendencies. For example, when a child feels threatened the kindness and affection of the parent can help them feel calm because their brains are set up to be calmed by the kindness of others (Ashworth et al., 2011). A key focus of CFT is that our relationships with ourselves can be helpful, kind, compassionate, understanding, and validating rather than self-critical (Gilbert & Irons, 2005).

The use of yoga in balancing the affect regulation systems

CFT suggests that the way we think and feel about ourselves and others is linked to the notion that there are three different types of emotion regulation systems (Depue & Morrone-Strupinsky, 2005). The first system is the threat system, which is evolved to detect threat and instigate defensive/protective actions. It is associated with emotions such as anger, anxiety and defensive behaviours such as fight, flight, and freeze responses (Panksepp, 1998). The second significant affect regulation system is focused on acquiring and achieving resources.

The purpose of this system is to give us positive feelings that energise and guide us to seek out things; it is a system of desires that guide us to important life goals (Depue & Morrone-Strupinsky, 2005). The third regulation system is related to contentment and being neither threatened nor driven to succeed. Contentment is associated with a sense of peacefulness, well-being, and quiescence (a state of not seeking). Contentment is not just the absence of threat or reduction of the stimulation of the threat protection system, it is the activation of a particular system linked to opiates that mediates feelings of well-being and contentment (Gilbert, 2010). This activation involves the stimulation of the opiate-oxytocin system. Oxytocin is a neurohormone linked to feelings of affiliation, trust, and feeling soothed and calmed within relationships (Depue & Morrone-Strupinsky, 2005). The CFT approach proposes that the soothing contentment system can be activated in various ways, such as by actions of others or practised imagery (Ashworth et al., 2011). Gilbert (2010) highlights that the main focus of CFT is in balancing the three affect regulation systems and this is partly achieved by activating the self soothing system.

While the CFT approach uses specific training exercises that help to promote compassionate attending through thinking behaviour and imagery, the use of yoga within a holistic rehabilitation setting may be a further method to help stimulate the self-soothing system. Rao and colleagues (2013) highlight that yoga may be a mechanism for the elevation of oxytocin levels. The elevation of oxytocin levels is associated with a sense of well-being.

Method

Client and clinical context: neurological symptoms and background

Karen suffered a sub arachnoid haemorrhage (SAH) at the age of thirty, three years prior to engaging in holistic rehabilitation. She provided consent to take part in this research to her psychologist. Some details have been altered to preserve anonymity.

Neuropsychological assessment and rehabilitation

Karen was assessed at a holistic neuropsychological rehabilitation centre three years after her SAH. A biopsychosocial approach underpinned a formulation that took account of the adverse emotional consequences of her brain injury in conjunction with other acquired difficulties and social participation restrictions (Williams & Evans, 2003; Wilson et al., 2009). Karen attended an eighteen week rehabilitation programme aimed at supporting her to (a) develop an awareness and understanding of the consequences of her brain injury, (b) adjust emotionally to these effects, and (c) develop strategies to manage her cognitive difficulties in the context of her main goals to successfully return to work and manage her emotions more effectively.

Psychological assessment and formulation

The initial assessment was aimed at formulating Karen's emotional difficulties within the context of her brain injury and related difficulties with social participation. Assessment involved clinical interviews with Karen and key family members, in addition to standardised questionnaires. Karen's emotional symptoms included feelings of anxiety, depression, and low self-esteem and self-criticism. Her history included childhood experiences of verbal abuse from family members. She experienced many highly self-critical thoughts in regard to her experiences and her abilities. Karen reported that she was very self-critical as she felt that her difficulties were her own fault. The CFT approach highlights that highly critical individuals can find it difficult to empathise with themselves or to emotionally nurture themselves through self-soothing. The stimulation of the threat system can be maintained through anger and frustration (Gilbert & Irons, 2005). The current study focuses on the use of a weekly yoga class at a holistic rehabilitation centre to help activate the self-soothing system in conjunction with psychological therapy.

The structure of the yoga group

The yoga group was facilitated by an experienced Hatha yoga teacher and took place weekly for sixty minutes at the rehabilitation centre. The group took place over a ten week period and was usually attended by three clients and three members of staff. The number of clients and staff in attendance at each session varied intermittently during this period. The sessions followed a consistent structure of dynamic warm ups, standing yoga poses, seated yoga poses, and deep release and relaxation. The group was aimed at any level of experience and members were encouraged to participate at their own pace. The facilitator promoted an atmosphere of trust and safety in order to help members to feel comfortable and the non-competitive philosophy of yoga was emphasised.

Client experiences of the yoga group

Karen was interviewed using a semi-structured interview for forty-five minutes after she had undertaken ten weekly yoga sessions. The questions were open-ended and were aimed at understanding her experiences of the yoga group. A thematic analysis of the transcription was undertaken using interpretative phenomenological analysis (IPA) (Smith et al., 1999). From this thematic analysis it emerged that the yoga group had helped her to develop techniques to help stimulate her self-soothing system. From the transcript the following themes were ascertained:

The relaxation and mindfulness go with you

Karen described how the yoga sessions helped her to practise relaxation and mindfulness techniques and how she was able to continue these outside the yoga sessions.

Help feeling self-soothed

Karen described how she felt soothed when she had the opportunity to take time out from her busy life and concentrate on her breathing and posture and found it calming. She described feeling self-soothed as result.

Helps with anxiety

Karen described how the yoga sessions enabled her to feel calm and reduced her feelings of anxiety. She described how she felt a sense of calmness after the sessions that helped her feel safe.

Focus on body and mind

Karen described how prior to undertaking holistic rehabilitation she did not see the value in taking time out for herself. She felt that undertaking yoga sessions each week has highlighted the importance of taking time out and focusing on her body and mind. This in turn has helped her to do more for herself and has had a significant impact on her ability to reach goals, as the yoga has assisted her with clarity of thought and ability for self-reflection. She felt it also promoted a sense that she deserves to have time out from daily activities and felt it increased her sense of well-being. Karen described being able to reflect on her goals in a non-threatening manner as the yoga sessions enabled her to learn how to relax and take a step away from every-thing that is going on around her. Karen also stated how taking time out by undertaking yoga was important in the context of undertaking intensive holistic rehabilitation as the nature of the yoga sessions meant that she was able to take time out from intensive sessions. She also reported that the time out in yoga sessions assisted her to put her goals in perspective and realise that they were achievable without the clouding of worrying thoughts.

Discussion

This case study illustrates that a weekly yoga group can help stimu-late the self-soothing system in conjunction with CFT in a holistic rehabilitation setting. The themes suggest that the yoga sessions can help promote self-soothing and self-soothing techniques, which are a key factor in CFT in helping to promote the stimulation of the affilia-tive and self-soothing system in order to reduce self-criticism. The themes also reflect the reduction of the threat system through reduc-ing feelings of anxiety and the promotion of a sense of well-being.

Based on Karen's experiences it is suggested that the downregula-tion of self-criticism and worry through the stimulation of the self-soothing system by attending a weekly yoga group enabled Karen to access her drive excitement system in a non-threatening way. Gracey and Ownsworth (2011) advocate working at the implicational level of meaning to support identity change, adjustment, and adoption of rehabilitation strategies through integrated rehabilitation techniques following brain injury. Karen's experiences suggest that the yoga group, through mindfulness techniques and the ability to take time out and opportunity for self-reflection has facilitated change at the more personally salient implicational level of meaning.

The themes reflect how attending the yoga group helped to promote the use of mindfulness and relaxation techniques in a safe environment. Both these techniques are of well-known utility for the reduction of anxiety. Karen also reflected that the non-threatening atmosphere of the group was useful when undertaking these techniques. An environment that promotes a sense of safety, trust, and cooperation is at the heart of a holistic rehabilitation programme (Wilson et al., 2009). The current case study suggests that a yoga group could be beneficial to enhancing the therapeutic environment through these shared values.

It is important to acknowledge that the findings of the study are based on a single case using a qualitative methodology that focuses on subjective experiences. The yoga intervention was conducted in the context of a holistic rehabilitation programme. Further case studies and multiple centre research that employ standardised outcome measures for well-being and for anxiety would be useful in order to increase theoretical understanding of how yoga groups may down-regulate the threat system and increase feelings of well-being. This could contribute to the broadening of knowledge for the use of yoga in neuropsychological rehabilitation. Further studies using a three group approach including a placebo control group may delineate the role yoga plays in stimulating the self-soothing system from the role of the CFT intervention. The current case study clearly illustrates that yoga could potentially enhance psychological interventions in a holistic neuropsychological rehabilitation setting.

Acknowledgements

Thanks to Karen, Doug, Katrina Dick, Jill Winegardner, all of the Oliver Zangwill team, and to Gill Olsen.

References

Ashworth, F., Gracey, F., & Gilbert, P. (2011). Compassion focused therapy after TBI: theoretical foundations and a case illustration. *Brain Impairment, 12*: 128–139.

Bastille, J. V., & Gill-Body, K. M. (2004). A yoga-based exercise program for people with post-stroke chromic hemiparesis. *Physical Therapy, 84*: 33–48.

Depue, R. A., & Morrone-Strupinsky, J. V. (2005). A neurobehavioral model of affiliative bonding. *Behavioural and Brain Sciences, 28*: 313–395.

Gilbert, P. (2010). Compassion focused therapy [Special Issue]. *International Journal of Cognitive Therapy, 3*(2): 95–210.

Gilbert, P., & Irons, C. (2005). Focused therapies and compassionate mind training for shame and self attacking. In: P. Gilbert (Ed.), *Compassion: Conceptualisations, Research and Use in Psychotherapy* (pp. 263–325). Hove: Routledge.

Gracey, F., & Ownsworth, T. (2011). The experience of self in the world: the personal and social contexts of identity change after brain injury. In: J. Jetten, C. Haslam, & A. Haslam (Eds.), *A Social Cure: Identity, Health and Wellbeing* (pp. 273–296). Hove: Psychology Press.

Lavey, R., Sherman, T., Muesser, K. T., Osborne, D. D., Currier, M., & Wolfe, R. (2005). The effects of yoga on mood in psychiatric inpatients. *Psychiatric Rehabilitation Journal, 28*(4): 399–402.

Lundgren, T., Dahl, J., Yardi, N., & Melin, L. (2008). Acceptance and commitment therapy and yoga for drug-refractory epilepsy: a randomized controlled trial. *Epilepsy & Behaviour, 13*(1): 102–108.

Lynton, H., Kligler, B., & Shiflett, S. (2007). Yoga in stroke rehabilitation: a systematic review and results of a pilot study. *Topics in Stroke Rehabilitation, 14*(4): 1–8.

Michalsen, A., Grossman, P., Acil, A., Langhorst, J., Ludtke, R., Esch, T., & Dobox, G. (2005). Rapid stress reduction and anxiolysis among distressed women as a consequence of a three month intensive yoga programme. *Medical Science Monitor, 12*: 555–561.

Panksepp, J. (1998). *Affective Neuroscience*. Oxford: Oxford University Press.

Rao, N. P., Varambally, S., & Gangadhar, B. N. (2013). Yoga school of thought and psychiatry: therapeutic potential. *Indian Journal of Psychiatry, 55*: 145–149.

Smith, C., Hancock, H., Blake-Mortimer, J., & Eckert, K. (2007). A randomised comparative trial of yoga and relaxation to reduce stress and anxiety. *Complementary Therapies in Medicine, 15*: 77–83.

Smith, J. A., Jarman, M., & Osborne, M. (1999). Doing interpretative phenomenological analysis. In: M. Murray & K. Chamberlain (Eds.), *Qualitative Health Psychology: Theories and Methods* (pp. 218–240). London: Sage.

Vancampfort, D., De Hert, M., Knapen, J., Wampers, M., Demunter, H., Deckx, S., Maurissen, K., & Probst, M. (2011). State anxiety, psychological stress and positive well being responses to yoga and aerobic exercise in people with schizophrenia: a pilot study. *Disability and Rehabilitation, 33*(8): 684–689.

Williams, W. H., & Evans, J. J. (2003). Brain injury and emotion: an overview to a special issue on biopsychosocial approaches in neurorehabilitation. *Neuropsychological Rehabilitation, 13*: 1–11.

Wilson, B. A., & Gracey, F. (2009). Towards a comprehensive model of neuropsychological rehabilitation. In: B. A. Wilson, F. Gracey, J. E. Evans, & A. Bateman (Eds.), *Neuropsychological Rehabilitation: Theories, Models, Therapy and Outcome* (pp. 1–21). Cambridge: Cambridge University Press.

Wilson, B. A., Gracey, F., Evans, J. E., & Bateman, A. (2009). *Neuropsychological Rehabilitation: Theories, Models, Therapy and Outcome.* Cambridge: Cambridge University Press.

CHAPTER SEVEN

Body in mind training: mindful movement for the clinical setting*

Tamara A. Russell and Tiago P. Tatton-Ramos

Part 1: theoretical underpinnings of body in mind training (BMT)

The task of therapy

For clients entering psychological services, a key task is to enhance emotion regulation skills (Berking et al., 2008). Emotion regulation is defined as "all of the conscious and non-conscious strategies we use to increase, maintain, or decrease one or more components of an emotional response" (Gross, 2001, p. 215). Emotion dysregulation presents in both chronic and acute forms and arises following emotional or physical trauma (Cicchetti et al., 1995; Thayer & Lane, 2000). Fundamentally, a proactive and effective emotion regulation requires the integration of bodily and cognitive processes in a way that brings into consciousness the sensations, mental habits, and emotional states that usually lie outside of awareness (Maiese, 2011; Mehling et al., 2011; Niedenthal, 2007; Thayer & Lane, 2000).

What is currently offered?

In the currently prevalent cognitive behavioural therapy (CBT) approach, the collaborative therapeutic endeavour requires access to

---

* Originally published in 2014 in *Neuro-Disability & Psychotherapy*, 2(1/2): 108–136.

and manipulation of dysfunctional cognitions and attitudes (Beck & Beck, 2011; Leahy, 2006), emphasising a more mental as compared to bodily psychotherapeutic dimension. This explicit process relies on the ability to overtly label and engage with cognitive and emotional material—changing cognition in order to impact on emotion (Leahy et al., 2011). While a large evidence base supports this approach (Roth & Fonagy, 2005) clients with neurological damage, learning difficulties, or with chronic, complex, or severe issues, may be less well served by this approach (Lynch et al., 2010; Meyer & Hautzinger, 2012). Psychoanalytic approaches work more implicitly (Schore, 2011) through slow and dynamic modifications of the underlying structure of mind. This makes the effects hard to measure and as a result, psychodynamic approaches are often positioned outside of the mainstream health services (Sandell et al., 1997).

More recently, the growing field of mindfulness based interventions (MBIs) offers a new and refreshing perspective (Eberth & Sedlmeier, 2012; Kabat-Zinn, 2003). These approaches include mindfulness-based stress reduction (MBSR), mindfulness-based cognitive therapy (MBCT), acceptance and commitment therapy (ACT), and dialectical behavioural therapy (DBT). All of these approaches integrate aspects of mindfulness (to a greater or lesser degree) and a growing evidence base supports their efficacy across a range of mental and physical health conditions (Baer, 2005; Carlson, 2012; Mace, 2007).

The MBIs are structured in an integrative body/mind framework, with most working through the body via a specific aspect of cognition (attention training). MBIs protocols support the development of "the moment by moment attention to bodily sensations and thoughts" (Kabat-Zinn, 2003; Segal et al., 2002; Williams & Penman, 2012). The progressive development of different metacognitive abilities is the driver behind this training, but the fundamental role of the body to the effects is often played down (Worsfold, 2009).

The body in mindfulness training

In Buddhist traditions, mindfulness starts with the breath and body.

> If one thing . . . is developed and cultivated, the body is calmed, the mind is calmed, discursive thoughts are quieted, and all wholesome

states that partake of supreme knowledge reach fullness of development. What is that one thing? It is mindfulness directed to the body. (Bodhi, 2012)

This emphasis on the body is not exclusive to Eastern philosophies. Nietzsche commented "Behind your thoughts and feelings, my brother, there stands a mighty ruler, an unknown sage—whose name is self. In your body he dwells; he is your body. There is more reason in your body than in your best wisdom." (Nietzsche, 1954).

Mindfulness of the body is the first step in the task of calming the mind and regulating emotions as taught in MBIs and a different sort of engagement with the non-verbal "felt sense" of the body is fundamental to MBIs and their effects. The fact that bodily sensations can only be experienced in the present moment, and particularly so during movement, makes them an accessible entry point to mindfulness. However, assessment of bodily awareness is not usually a feature of these studies. Measurement of bodily awareness is a developing, but problematic area of research (Mehling et al., 2011), relying on subjective reports of non-verbal experiences.

Neurophysiological research demonstrates that mindfulness does indeed, start with the body (Kerr et al., 2013). Kerr and colleagues state that,

Learning to control alpha oscillations in SI [primary somatosensory cortex] through localized body-focused attention may be a key gateway mechanism for learning to use thalamocortical alpha regulation to suppress irrelevant sensory inputs across sensory neocortex in an internally directed, top-down manner, for forms of regulation such as selective attention and working memory. (Kerr et al., 2013, p. 12)

From attending to the body in this way, follows the ability to regulate more abstract mental sensations such as thoughts, memories, images etc.

Evidence from MBIs supports this notion, if indirectly. The mindful yoga component in MBSR seems particularly potent. The time of yoga homework practise was significantly correlated with a large number of outcome variables: despite being completed on fewer days, and for shorter duration (Carmody & Baer, 2008). It was the only practice significantly correlated with changes in the "non-judging" component of the five factor mindfulness questionnaire (FFMQ). A predominantly yoga-based mindfulness protocol evaluated in (one of the few) randomised controlled trials (RCTs) was shown to be helpful in binge eating disorder (McIver et al., 2009).

Tang and colleagues created an integrative body–mind protocol (based on Chinese medical theory). They showed that after eleven hours of training, healthy participants improved their performance on an attention test. There were also changes in the functional activation and white matter tract integrity of the anterior cingulate cortex (Tang et al., 2012). These studies suggest that body-based practices have a particular potency that may currently be minimised. Worsfold (2009) reminds us that mindfulness of the body as taught in MBIs is more than just a vehicle to effect change in metacognition and highlights the lack of discussion in the mindfulness literature, and clinical theory in general, regarding how the body is conceptualised (Worsfold, 2009).

One speculation is that as many MBIs have been adopted by mental health workers they have migrated into the more known "mental" domain. This may be due to a training that prioritises the mental, lack of confidence in body-based practices, or the prevalence of burnout and "wounded healers" in the profession (Carl Jung and Marsha Linehan being two well-known examples) making it hard for this group to engage with their own bodily experience. Although MBIs are offered to our clients, the suffering and need among staff is significant. Staff in UK mental health teams had the highest rates of depersonalisation and emotional exhaustion in a recent European comparison (Hill et al., 2006).

It seems there is much potential for more body and movement ways of working and this is currently underdeveloped. It is essential to understand this aspect of mindfulness is not just a stage to pass through. In some traditions such as Daoism, mindful movement is taken to the highest degree of spiritual endeavour (Liao, 2000).

Other body-based practices

A growing number of body psychotherapies and somatic education methods (Fogel, 2013; Totton, 2005) are attracting the attention of mainstream health providers, particularly for clients who struggle to engage with mainstream therapies. These include Feldenkrais, Rolfing, Eutonia, and Continuum Movement to name a few (Totton,

2005). Anecdotal and case study reports are favourable, but an issue with this work is the compartmentalisation of the different traditions making it hard to discern a cohesive underlying mechanism of action. The evidence base, perhaps for this reason, remains sparse (Davis, 2009; Ives, 2003).

Two body-based practices with a long history and theoretical foundation (non-biomedical) are the Eastern mind-body practices of t'ai chi and yoga. These terms broadly describe a range of practices with a growing evidence base for efficacy across a range of physical (Buffart et al., 2012; Lan et al., 2013; Liu et al., 2010) and mental (Cabral et al., 2011; Wang et al., 2010) conditions. The predominant explanation for their efficacy rests on the consequences of relaxation response activation (Dusek & Benson, 2009; Jahnke et al., 2010). However, this is only one part of their mechanism and likely selling them far short of their healing potential.

A challenge with appraising these approaches is their mode of action/learning. In movement practices such as t'ai chi, an implicit learning of internal patterns of body and mind occurs over years of practise and repetition of the slow and gentle physical movements. Hayes and Broadbent (1988) have defined implicit learning as "the unselective and passive aggregation of information about the co-occurrence of environmental and features". This type of training makes the ability to report on what has changed (and how) limited as it has entered the "system" in a non-verbal, non-conceptual way (Liao & Masters, 2001). This is supported by Carmody and Baer's (2008) observation that yoga practice in MBSR was correlated with all except the "describing" factor of the FFMQ.

This way of learning is reflected in a pedagogy that uses observation, mirroring, repetition, tactile adjustments, and little overt instruction. The classic texts of t'ai chi (Liao, 2000) can increase confusion as the language describing the concepts makes much use of metaphor. The implicit learning and use of metaphorical language can be difficult for general Western audiences and make it hard for empirical studies to be conducted (Sieh & Ralston, 1993).

Body in mind training

BMT works in an optimum and still relatively unexplored zone, between the underdeveloped opportunities for mindful movement as a part of MBIs and making more explicit the ancient body expertise from t'ai chi and martial arts. Using mindful attention to the body in an innovative way (Chiesa et al., 2013; Kerr et al., 2013), BMT aims to mine the edge between non-conscious (implicit) and conscious (explicit) bodily, cognitive, and emotional experiences. It works from a "bottom-up" body-oriented stance with the bodily sensations (including those associated with emotional states) as the object of the "top-down" attention (Brewer et al., 2011; Chiesa et al., 2013; Russell, 2011). The aim was to develop mindful movement exercises for use by anyone who might struggle to engage with the currently available MBIs, thus widening access to mindfulness. A secondary aim was to help clinicians begin to work in a mindful way using the body, so that they too can benefit (Russell, 2011).

As a foundation—especially for the design and delivery of the exercises—BMT uses the growing neuroscientific knowledge base around how the body is represented in the brain. Additionally, embodied learning principles are inherent in the teaching (Bohannon, 2010; Bresler, 2004) creating a learning experience that is embedded at a deeper level. In the following sections, the developmental process and underlying scientific rationale are detailed, followed by some reflections on the clinical implementation.

What is so important about movement?

Even the most basic notion of the "self"—the conscious being—depends on the movement of an organism (Damasio, 2012; Dehaene & Naccache, 2001). When we begin to move independently in the world a "psychological revolution" occurs as multiple cognitive and social processing abilities come on-line (Anderson et al., 2013). Functional activity of the growing infant plays a key role in the "formation, construction and development of structure in the nervous system" (Anderson et al., 2013, p. 11). Key structures of the brain that activate the organism to engage with the environment and others, including basal ganglia and cortical motor loops, have a role in cognition (Middleton & Strick, 2002). Wolpert and colleagues (2011) have even gone so far as to suggest that movement was the real reason brains evolved (see his TED talk entitled "The real reason for brains"). The development of brain regions controlling motoric behaviour may thus underpin the subsequent evolution of "higher-order" regions sub-serving emotion and cognition. Within cognitive science, a growing research field, embodied cognitive neuroscience, considers more deeply the role of the body in cognition (Gallagher, 2006).

Movement and emotion: approaching and avoiding

Davidson's (1992) update on early theories of emotion (Ekman, 1992; Tomkins, 2008) highlights the functional role of emotions, driving the organism to either approach or avoid evolutionarily important stimuli. Our movements therefore can provide clues as to our underlying emotional and motivational state. Gray's neuropsychological theory of anxiety (Gray & McNaughton, 2000), based on extensive animal work, details a behavioural approach system, a fight/flight system, and a behavioural inhibition system. Similarly, Cloninger has proposed a psychobiological model of temperament and character that captures the fundamental integration of motor, cognitive, and emotional inhibition patterns. These play out in our lived experience and personality, including our propensity for novelty seeking and harm avoidance behaviours (Cloninger et al., 1993; Gardini et al., 2009). Emotional and motivational states are thus intimately linked with movement.

Movement and mood are closely related

In many psychiatric conditions, disruptions in approach and avoidance movements are evident; reduced activation in depression, physical avoidance in anxiety, agitation in mania and psychosis (Noggle & Dean, 2012). Not moving the body impacts on mood, as seen in neurological conditions that affect mobility (such as Parkinson's) where there is high co-morbidity with depression (Cummings, 1992). Mild and severe mental distress can be modified by movement/exercise (NICE, 2004) and in healthy individuals exercise improves cognitive performance (Chaddock et al., 2012; Hillman et al., 2008) in a way observable in the brain (Chapman et al., 2013; Nithianantharajah & Hannan, 2009). Researchers such as Meijer (1989), Wallbott (1988), and, more recently, App et al. (2011) suggest that movements help us to not only express but also process emotions, with movements providing information that can help us recognise what we are feeling.

Summary

In summary, movement and the motor system provide an essential foundation for our emotional and cognitive lives. Fundamental movements towards or away from things are enacted by the body, in the living world (Candidi et al., 2012; Gibbs, 2006; Maiese, 2011). Movement and mood are intimately linked; lack of movement impacts negatively on us, and movement, positively. BMT makes movement the central training tool for these reasons. Body-based practices such as t'ai chi bring a wealth of knowledge about the moving body—when this is made explicit, with mindful intention and attention, possibilities for change in the whole mind–body system occur. In the next section, the development process of the BMT program is described.

Part II: the development process of BMT

BMT arose from the integration of three disciplines: neuroscience, clinical psychology, and martial arts. In the following sections the developmental process to bring these exercises into the mainstream health setting are briefly detailed. The reader is referred to Russell (2011) and a forthcoming book for more details (Russell, 2015).

Phase 1–BMT in the acute psychiatric setting

Working in the adult psychiatric mental health setting and predominantly with those experiencing severe and enduring conditions, a series of mindful movement exercises were piloted across in and outpatient settings (Russell, 2011). Table 1 provides a summary of the activities conducted. The purpose of these groups was to determine the acceptability and feasibility of the exercises and gain feedback from participants about what they found helpful in the class and what they utilised in daily life. The groups were predominantly offered on male wards to understand whether this approach, based on martial arts and neuroscience might appeal to this group. In the mindfulness literature, females are twice as likely to remain engaged with MBIs (Kabat-Zinn & Chapman-Waldrop, 1988), and there are predominantly female participants in MBCT studies (Piet & Hougaard, 2011; 63–81% of the samples in their meta-analysis were female). Engagement with "meditation" or "yoga" may be a barrier for males.

Table 1: Activities in the BMT class with the links to mindfulness theory and t'ai chi theory

Basic warm up: slow, intentional, mindful movements of the major joints (neck, wrists, shoulders, elbows, waist, hips, knees, ankles). Exercises completed to the ability of the individual, with modifications suggested as required.

Mindfulness theory link: emphasis on focusing and maintaining attention to bodily sensations and listening to the body to find your own pace/limit (compassion). Experiments with pacing to illustrate how slowing down (or pausing) allows us to see more and guidance on tracking intention to move prior to movement.

T'ai chi theory link: highlighting transitions from stillness to movement and back to stillness, exploration of effort and ease in the movements, increasing range of motion (stretch) via integration of bodily sensations, mind, and breath.

Stationary elements: working left/right brain with co-ordinated hand and leg movements (e.g., the preparation for the t'ai chi movement "ward off" and static "wave hands like clouds").

Mindfulness theory link: mindful awareness of mental reactivity and the habits triggered when challenged with a novel movement sequence. Highlighting differential sensory input from left and right sides of the body to pique curiosity and increase self-directed exploration of the body.

T'ai chi theory link: breaking down movements into segments to re-build in a more integrated way, attending with focused and broad attention to the whole body during a movement, working with the principle of *fang sung* (a state of alert relaxation in the body).

continued

Table 1: Activities in the BMT class with the links to mindfulness theory and t'ai chi theory (*continued*)

Dynamic (moving) elements: working balance, posture, co-ordination, and weightshifting, for example, "wave hands like clouds" and "repulse the monkey". Waist turning practices and sequences that emphasise the sequential and consequential connections between body parts and movement.

Mindfulness theory link: focused and broad attention highlighted throughout the movements attending to proprioceptive and kinaesthetic feedback, awareness of intention and execution phases of the movement, alertness to mental habits and mind-wandering throughout the movement, awareness of the temporal sequence of movements of body (and mind)

T'ai chi theory link: using the mind (*yi*) to move the body, more complex breaking down of movement sequences in order to rebuild, *fang sung* while moving.

Walking meditation: controlling weight, posture, and movement. Noting and not reacting to mind wandering during the walking, observing the mental states of restlessness/agitation and dullness/boredom. Emphasis on exploring effort and ease throughout the movement, and *fang sung*.

Note: a more detailed description of these exercises can be found in Russell (2015). Training the BMT is available via www.mindbodymot.com and on line training is forthcoming.

The t'ai chi exercises were predominantly drawn from the Shibashi (Set One) developed by Lin Housheng (www.linhousheng.com/) a leading qigong practitioner. This set combines contemporary yang style t'ai chi movements and qigong exercises. The ten principles of t'ai chi by Chen-fu (Wainapel & Fast, 2003) were embedded in the teaching (see Table 2). Participants are guided to attend to these aspects in an explicit and mindful way. These principles are fundamentally related to developing a deeper awareness of the body and posture during movement and stillness.

Ward staff took part in many of the classes and feedback was obtained about their reactions and how they felt clients had engaged. Primarily they were surprised how engaged participants remained in the one hour class, commented on a better understanding of "psychology in action", and observed their own need for this type of activity. The qualitative outcomes from this work indicated that for both clients and staff, BMT mindful movements provided a relaxing and engaging experience that increased self-mastery and body awareness (Russell, 2011).

Phase 2—BMT framework for health professionals

In Phase Two, BMT exercises formed the basis of workshops and short trainings for health care staff. Workshops were delivered nationally

Table 2: Ten principles of t'ai chi embedded in all BMT exercises

(i)	Keep the head and neck straight
(ii)	Upper and lower back kept in a straight line with the pelvis tucked under, softness in the knees.
(iii)	Separation of waist and hips, loosening the hips and groin.
(iv)	Shoulders and elbows are relaxed and down.
(v)	Upper and lower parts of the body move as one unit.
(vi)	Differentiating between a full (solid) and an empty (awareness of weight distribution).
(vii)	Moving with awareness of mind intention and minimal external muscle force.
(viii)	Transition between movements in a smooth, continuous manner.
(ix)	Assure a sense of harmony between the internal and external body feeling.
(x)	Experience a tranquil, meditative state, breathing in a smooth, continuous manner.

(UK) and internationally (Brazil, Barbados, Poland, Turkey) to a variety of health care workers and educators. In the UK, they were predominantly delivered to mental health multi-disciplinary team members (doctors, psychologists, nurses, health care assistance, physiotherapists, social workers, and occupational therapists) providing support for those severe and enduring conditions (Community Mental Health Teams (CMHTs), learning disability services, eating disorders).

The BMT Framework was developed as a way to unpack Kabat-Zinn's definition of mindfulness—"the awareness that arises from paying attention, on purpose, moment by moment and non-judgementally" (Kabat-Zinn, 1982, 2003) and allow those who were curious about mindfulness but without formal training to explain mindfulness to a carer, colleague, or client and begin to experiment with the principles of mindfulness in their lives (Figure 1). Using this definition and the BMT exercises as the experiential component, five key principles are explored; (i) pausing (as a means to access the present moment); (ii) intentionality (capturing the element of "on purpose"); (iii) attention (understanding what captures our attention and how we can train voluntary attention); (iv) observation (capturing the element of a different sort of relationship to experience); and (v) compassion (capturing the element of the non-judgemental attitude) (see Figure 1).

The understanding of the rationale and role of the mindful movement exercises was enhanced by providing conceptual information about the key theoretical principles of mindfulness and the relevant neuroscience findings. This aspect is vital to the BMT approach, directly drawing on what we know about the brain to deliver and explain the training. Working using a neuroscience framework provides an opening to discuss mindfulness in contexts more used to working within the biomedical model (e.g., mainstream health settings). It widens access for both staff and clients, so more people might benefit from this training.

The BMT framework is a means for individuals to begin to explore how they could bring mindfulness into their personal and professional lives. There was an explicit intention that the BMT exercises are as necessary for these staff as they are for clients. The workshop was considered a success if staff members commented "I need this too".

Written feedback was obtained from a number of these workshops and independently audited by a qualitative researcher with experience working in the complementary health sector (Wilkinson, 2013).

When can you slow down, or pause?
How will you remind yourself to do that?
Can you pause during events that are
pleasant, unpleasant, or neutral—what do
you notice?

What can you do to be more aware of
intention? When and where can you check in
with your intentions? How can you really
take a look at what is occurring right now in
your experience?

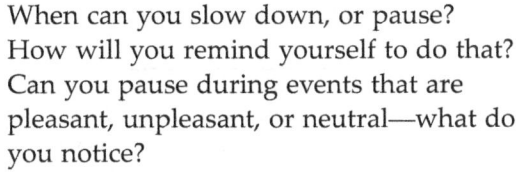

How can you remind yourself to notice where
your attention is?
What draws it away? How does it come back?
Is the focus narrow or wide?

What would it be like to really observe and
study mental and physical phenomena—like
a scientist?

How can you be kinder to yourself and/or
others and more at ease?

Figure 1: Body in mind training framework.

Using an adapted "framework" methodology (Huberman & Miles,
2002), emergent themes were derived and are shown in Table 3.
Further comments can be found in Appendix I.

While running these workshops the core training information and
exercises continued to be refined, based on the feedback and observation of what worked, what was clear, and where there was confusion.

Table 3: Key learning outcomes from BMT staff training workshops (N = 140 included in audit)

Primary Theme:	Mindfulness skill acquisition (using the BMT Framework) and developing an empathic and compassionate approach to the work.
Secondary Themes:	A new perspective for clinical work and improving professional practice.
	Incremental application of mindfulness within clinical practice.
	Understanding the need for sustained self-practice in preparation for clinical work.
	Potential application for specific client groups.
	Enhancing direct working with clients.,
	BMT as a way of building on previous mindfulness training and clinical applications of mindfulness.

From this emerged a series of exercises and conceptual teaching materials that clearly described the principles of mindfulness using mindful movement as the main training methodology. The materials were optimised to ensure individuals from a variety of professional trainings can work with the BMT framework across a range of clinical and non-clinical settings.

Phase Three—BMT-Five

From these workshops and the BMT framework (Figure 1) the BMT-Five session group programme emerged. This was an expansion of the principles of the BMT framework into five two and a half hour teaching sessions, exploring pause, intention, attention, observation, and compassion in turn. In order to provide a comprehensive learning experience with the intention of piquing curiosity and encouraging self-motivated engagement with the practices, each theme is explored via (i) physical mindful movements illustrating the main learning points; (ii) discussion of the underlying neuroanatomy; (iii) an explanation of the rationale; and (iv) discussion of real life application. Each theme is associated with a visual image, two of which (pause and free hugs (compassion)) are given as stickers to serve as a prompt in day to day activities (Figure 1).

In BMT, homework is optional and participants invited to implement some aspect of each principle into daily life, in whatever way is helpful or appropriate for them. For example, for some the pause might be a twenty minute body scan every day, while for others it might be taking a moment to be mindful of the soles of the feet before boarding a bus. From small steps and a more detailed conceptual understanding, the possibility for intrinsic motivation arises.

Six pilot groups with healthy individuals (including many health care staff) and individuals with bipolar illness were run in 2013, obtaining qualitative feedback. Two post-graduate dissertations will formally evaluate BMT-Five in 2014 and include measures of body awareness that evaluate mindful engagement with bodily sensations and the felt sense of emotion in the body (Mehling et al., 2012). It is anticipated that BMT-Five could provide an alternative to standard trainings, a pre- or post-MBSR or MBCT option, and/or a way for mindfulness teachers and health care providers to enhance their clinical mindfulness offering and care for themselves in their work.

Summary

In summary, the BMT exercises have evolved from early beginnings as a method to work using mindful movement for those who are very distressed into a structured training protocol based on mindful movement. It has been through the "filter" of a large number of health care staff who recognise both their own need for these practices but also the benefits and different entry point the BMT approach might bring to their clients. This predominantly qualitative data obtained in the development process will be augmented by future studies using quantitative methods.

Part III: BMT in action

Mindful movement (MM) is defined in BMT as any movement conducted with full explicit awareness of intention, attention, and all the physical and mental sensations unfolding over time. Mindful movements are conducted with a stance of compassionate acceptance towards each and every experience including thoughts, feelings, memories, and emotions but especially bodily sensations. The following section provides a description of BMT training principles and how they are applied in the clinical setting. To be clear, although BMT includes certain exercises that form the basis of the current five week protocol, it is also a set of guiding principles (Figure 1), allowing a flexible delivery that can meet the needs of the client group or setting.

Pausing and inhibition

Pause is the first guiding principle of BMT. Various movements are explored at a progressively detailed level with curiosity in a slow and gentle manner. Not only does this reduce the possibility of injuries, but it also allows many more of the rich sensations from movement to enter awareness. Pacing as a general theme (in our movements, in our lives) is explored. The completion of slow MM activates the motor inhibition system (fronto-basal ganglia networks) to be engaged (Dillon & Pizzagalli, 2007). As many participants state "going slow is hard". Specific brain regions are linked to motor (fronto-basal ganglia networks), cognitive (orbitofrontal cortex), and top down regulation of emotions (ventromedial prefrontal-amgydala interactions), but a common region (right ventrolateral prefrontal cortex) subserves inhibition in a domain general way (Aron et al., 2004).

This raises the possibility that by training the motor inhibition area through slow MM, the domain general area is boosted and has a greater resource when called on by other domains. This may explain the emotion regulation benefits of t'ai chi (Hong, 2008) and rate of development—these practitioners are not overtly training in emotion or cognitive regulation, but they are indirectly supporting these processes working through the motor domain.

Slow movements of any sort illustrate directly how we can see more of our experience by reducing the pace. A whole body or individual body part movement can be used. An example of a stationary large limb movement might be "swimming backwards" (a single arm movement—swimming backwards). Pacing could just as effectively be explored in the seated position by lifting one little finger up and down slowly. The principle rather than the precise movement is important. This makes this approach suitable for a wide range of clinical populations including those with neuro-disabilities who may have difficulties with standing and/or walking.

The sensory consequences of movement

Moving the body creates psychophysiological changes (Anderson et al., 2013; Damasio, 2012). Depending on the individual, slow MMs may give rise to changes in cardiovascular or respiratory rates (Caldwell et al., 2009). Movement of the joints will provoke a change in the muscles, usually relaxation. This is important for two reasons. First, these physiological changes provide a concrete object for attention training, and second, they can generate insights. BMT participants comment on increased awareness of stiffness (which can arise from medication and inactivity) and a new experience of relaxation in the neck and shoulders. This generates insights such as "I didn't realise I was so stiff" or "I need to move more". Self-efficacy arises from these exercises as participants realise they can do something themselves to alter their physical and mental state.

The moving body is a particularly salient attention object

Attending to the body generally supports a present moment focus as bodily sensations cannot be experienced in the past or future. The moving body provides a greater wealth of sensations to observe (see paragraph above) and the movement itself can be the object of attention. Movements have a temporal sequence, unfolding over time in a way that supports the ability to keep attention on a constantly updated "now".

As mindfulness develops, sensitivity to the temporal aspects of the movement increases, and it becomes possible to explore separately the intention to move, the execution of the movement, and the sensory consequences of the movement (Kerr et al., 2013; Tang, 2011). In BMT developing this temporal sensitivity via movement practices provides a scaffold to learning how other experiences, including mental events (chains of thoughts) and emotions unfold over time.

Movement of the body and the resultant sensations are a strong stimulus for the attention network, making these exercises suitable for those with very busy or disturbed states of mind as mind-wandering is reduced (Tang & Posner, 2009). Additionally, busy clinicians might incorporate MM into their working life. Much can be observed about our underlying approach/avoid tendencies when we become mindful of how we move about the clinical environment.

The movement itself can be variable—a finger, one or both hand(s), a limb, the torso, or head—and adapted to the individual's physical ability. This means that while there are specific movements that really work best to illuminate mindfulness principles, in fact any movement can be conducted mindfully following the BMT framework.

MMs also provide the opportunity to explore the experience of transitions between movement and stillness, a central training principle in t'ai chi (La Forge, 2005). Observing how our intention to terminate or create a movement unfolds illuminates how intention underpins all motoric output and indeed all our actions in the world. Learning to detect change in behaviour patterns is helpful when we are seeking to change unhelpful habits (of body or mind).

Which body in mind?

Neuroscience knowledge informs both the design and delivering of the BMT intervention (Russell, 2011). The neural circuitry for the body and motor system are well mapped relative to the circuits for emotion and cognition (Haggard, 2005; Meltzoff, 1990). Early sensory processing of bodily sensations takes place in the somatosensory cortex, with further refinement of the signal as it passes through sensory and motor association cortices. Sub-cortical nuclei also contribute to the experience of the body in mind (Longo et al., 2010). The "extended body" in the brain is likely mediated by activation in the posterior parietal cortex and the "emotional body in brain" encoded in the region of the insula (Berlucchi & Aglioti, 1997; Longo et al., 2010).

The hierarchical layers of body representation in the brain are exploited in BMT. Exercises explore the difference between somato-sensation (raw sensations of the body in primary somatosensory cortex), somatoperception (the conceptual body encoded in various regions including posterior parietal lobe), and somatorepresentation (Longo et al., 2010). Exercises deliberately provoke conceptual and visual images of the body and contrast these to the direct sensations or tactile representations. These exercises illustrate just how much "post production" the brain does (and how quickly and unquestioningly we believe it to be "real"). Later this is linked with how we process emotional states in the body and a call to remain curious even when we think we "know".

Thus the brain has a network of regions that progressively create a representation of the body in the brain from the raw primary sensations in somatosensory cortex to the schematic representation of the body in the mind in the parietal lobe (Azañón & Haggard, 2009). In many disorders of body image seen in the clinical setting, it likely that a reliance on processing from the schematic body underpins some of the observed difficulties (Fuchs & Schlimme, 2009). Similarly, if you are disconnected from the body, or depersonalised it is likely some aspect of this network is operating at a lower metabolic level (Simeon et al., 2000). In the neuro-disability field there may be potential for MM to help those with phantom limb pain.

Maximising learning

It is easier to pay attention if the stimulus is interesting and the signal is strong. The neuroscience of the body representation in the brain is used to support the learning in two ways. First, we know that there are disproportionately large areas of the brain dedicated to processing signals from the hands and the face in the somatosensory and motor cortices (Kerr et al., 2005). These are therefore, particularly potent objects for mindfulness training making them suitable starting points for those who have difficulties with attention, those with very chaotic minds, or for use in distracting environments. Mindfulness of the face can also support training in social skills.

Second, most individuals have a hand preference and use-dependent cortical plasticity impacts on sensorimotor cortical representation (Hammond, 2002; Schwenkreis et al., 2007). Thus sensations on the left and right sides (and particularly in the hands) can vary greatly. BMT exercises use these neuroscience facts and the experiential experience of left and right to pique curiosity and help people really take a look at their experience and increase bodily awareness. This might be particularly relevant for those with hemi-paresis or weakness on one side of the body. Using the BMT approach the clinician might guide a mindful exploration of the difference in sensations, the ability to perform a movement, and fatigue points on the left and right side, noting both the bodily sensations and the mental sensations (reactivity, judging, comparing, etc).

Hand movements that activate the left and right motor cortices preferentially activate approach or avoidance behaviours (Brookshire

& Casasanto, 2012). This field of research raises interesting questions about how movements might be used to change problematic approaching/avoiding in clinical populations. As described above, these approach/avoid behaviours are often linked to emotional states.

Observing emotions

Engagement with movement and bodily sensations provides the foundation to develop curiosity and confidence in being in the body. This is later developed to support direct engagement with affect in the body. The field of affective neuroscience has uncovered how emotions are represented in the brain and the complex bidirectional nature of top-down and bottom-up processing on our emotional lives and behaviour (LeDoux, 1996; Panksepp, 1998; Vytal et al., 2013; Wager et al., 2007). Damasio and colleagues (2000) make a distinction between emotions (a physiological response including sensations, autonomic and somatic, muscular changes, movement, etc. mostly implicit) and feelings (a label that describes the summation of body states as experienced by the mind).

In the BMT the MM training intends to help differentiate the mind-body experience of emotions and feelings, learning to detect and engage with emotions as they move through the body. This experience is captured by the phrase "That really moved me" used in response to art or music. These experiences are not quite the same as the sensations experienced in the somatosensory or motor cortices when we move the body, but instead are the moving, throbbing, piercing, spiky sensations often described using metaphor and analogy. They are distinct from feelings that are the summation of the physiological experience of the emotion and additional mental activity (drawing up memories, labelling, etc).

In BMT the critical question is "What is the difference between feeling sad and attending to the raw sensations of the body that you have labelled as 'sad'?" Staying directly with the raw sensations increases emotional intelligence by developing sensitivity to emotional shifts and keeping the attention in the body reduces the engagement of mental avoidance strategies (body sensations are only ever in the present moment) (Fogel, 2013; Kerr et al., 2013; Tang, 2011). Working with individuals with bipolar, they became able to distinguish the subtleties of a sensation bubbling up from the stomach that was a happy feeling, as compared to on that had an underlying anxious edge, perhaps suggesting hypomania.

With practise, and indeed exposure to previously feared emotional states, confidence grows in the ability to tolerate even very strong emotions without trying to alter them or engage with them using the "thinking" mode. This type of emotional re-education may help to "reset" maladaptive and toxic processes of coping with emotions (Kashdan et al., 2006; Ruths, 2011).

BMT uses predominantly moving practices as the route to the body, which makes it more suitable for those experiencing particularly distressing mental phenomena. The movement provides a moment by moment updated mental target for the attention in the form of a wealth of sensations entering in the body. This provides a safe early engagement with the body that can later be used in the exploration of emotions and emotional reactivity. Training in movement detection also increases sensitivity to how things unfold over time, vital when we then enter the mental and emotional realms. "This too will pass" is a common phrase that supports this understand but working with MM provides an embodied training in this concept (Kerr et al., 2013; Tang & Posner, 2009).

The body and compassion

If we pay attention, we can notice very clearly the signals from the body telling us it is suffering, and we can do more than we think to be kind to ourselves (Tang, 2011). The mild range sensations of pain and/or discomfort that naturally emerge from the movement training posits us a challenge that provides a real chance to observe how we react in the mental/emotional domains.

BMT exercises explore this, using the concrete signals of the body and alertness to mental reactivity to illustrate what happens when we meet an experience we do not like and want to change or avoid. In a single movement there might be a chance to experience the sensations of (and reaction to) pleasant, unpleasant, or neutral experiences. In those with disabilities, there is an invitation to attempt the movement (or adapted movement) as best as is possible, noting whatever mental or physical sensations arise, including those related to struggle, non-acceptance, and the desire for things to be different (as they were before, like others, or how they might be in some imagined future, etc.).

A key principle of BMT training (and in mindfulness generally) is an invitation to try what is currently possible and mindfully explore the limits of one's ability. Observing, with curiosity and compassion what arises (in moments of "success" or "failure") both at the level of physical sensations and mental reactivity. Meeting fatigue (either mental or physical) is part of the training process. Observing mental reactivity and the psychological style when meeting limits is integral to therapeutic work moving clients towards acceptance of injury or illness.

Learning how we cope when things are not as we wish them to be is also highly relevant for staff (Hayes et al., 1996). Being compassionate in our work is difficult if we deny, suppress, avoid, or cut off from our own emotional experience and struggle. Training in mindfulness can increase self-compassion (Germer, 2009; Gilbert & Procter, 2006; Pace et al., 2012) and this training is as vital for staff (Shapiro et al., 2007) as it is for our clients (Van Dam et al., 2011). Some programmes such as the Mindful Self-Compassion Program prioritise this compassion aspect (Neff & Germer, 2013). Appendix II provides some guidelines as to how elements of the BMT Framework can be used by therapists in their work to support this process.

Imaging studies with experienced meditation practitioners have shown that when engaging in compassionate meditation, and provoked by (negative) emotional stimuli, there is greater activity in the insular cortex, amygdala, temporal parietal junction, and posterior parietal cortex relative to controls (Lutz et al., 2008). This suggests that the experience of the emotion may be enhanced when in a compassionate state. The ability to stay with the experience of strong emotions without reacting may be the signature of true compassion—both for our own or another's pain.

Increases in cortical gyrification have been observed in the right anterior insula (Luders et al., 2012), in a manner related to duration of practise and the right insular cortex becomes thicker (Lazar et al., 2005) and denser (Holzel et al., 2007) after meditation practise. The insular cortex codes interoceptive information related to the felt sense (including in the body) of emotion (Craig, 2005; Longo et al., 2010). It is ideally placed and connected to structures that allow it to monitor the internal state of the organism and co-ordinate other regions to allocate attention, evaluate the context and plan appropriate approach or avoidance actions.

The right insula is a brain region enhanced by mindfulness practise and involved in experiencing and regulating felt emotions in the body. Similar findings have been observed in those who have undergone shorter mindfulness trainings (Farb et al., 2007; Gard et al., 2012). Mindfulness training may therefore serve to enhance our emotion regulation abilities but paradoxically this is by allowing us to feel more, representing increased bottom-up processing of the stimulus (Holzel et al., 2011, p. 542).

Moving the body and mindful engagement with the movement of emotion in/through the body thus provides us with the possibility of developing compassionate states of mind. Compassion is linked to the ability to fully be present and open to all aspects of emotion, including the bodily component of the experience.

Summary and conclusions

In this chapter a preliminary framework for a new MBI has been presented. The BMT approach blends neuroscience with the principles of mindfulness and is presented as a framework that can be used by a range of clinical staff members across a range of clinical populations. Using this brain-based approach, clinicians with experience in their specialist areas and who have tried out the exercise themselves, should be able to create thoughtful adaptations allowing them to work with MM with a range of clients. The choice of movement and the depth of work can be informed by the clinician's experience and collaborative exploration with the client. The development work shows that staff from a variety of backgrounds can work with this approach both for themselves and with clients. Both staff and service users indicated that they found the addition of the basic neuroscience understanding informative and motivating. This is pivotal tool to enhance curiosity and encourage people to explore their experience in this more embodied way. As with all MBIs, the practice and experience of the clinician offering the training is important so those willing to try this way of working are strongly encouraged to engage as much as they can with their own body in mind. Some materials to support this are provided in Appendix II. It is hoped these ideas may stimulate debate and further research into this relatively new and exciting area.

Appendix I:

Selected staff comments from participants on the BMT Module 1 mindfulness training for health care professionals

A new perspective for my work.

This training will change my practice.

Now I feel better able to explore different aspects of mindfulness for myself and my clients.

I liked this different approach to the mindfulness training working so much with the body.

This should be mandatory for all staff.

I feel I can bring some "mini" practices into my working day and feel more confident now about using it working with others.

The workshop was very engaging and the exercises were pitched at the right level. A really useful introduction to mindfulness.

I loved the playful and humorous attitude—this made me more confident to experiment with mindfulness for myself.

I was really shocked at how much I could noticed when I slowed down.

Appendix II:

Suggestions for using the BMT framework as therapist

This framework can be used on a moment by moment basis within the session in order to ensure, as far as possible, that mindfulness is maintained and embodied by the therapist. Some illustrative applications are provided in Table 4 below. The framework also provides a helpful reflective tool post session or for supervision, checking in with each aspect and reflecting on whether this was apparent in the session and thinking about where it might be enhanced or noticed in future interactions.

Acknowledgments

Tiago P Tatton-Ramos is supported by a doctoral fellowship from the CAPES Foundation—Brazil

Table 4: BMT framework for therapists in the session

Pause	Pausing before replying or interpreting a client's comment.
	Allowing silences and in that time noting any bodily reactions.
	Transitioning between clients or clinical tasks.
	Reflected on during times when we notice we want to "do more" both in body and mind for a client{emdash}often a signal that we are seeking to reassure our own anxieties rather than theirs.
Intention	Extremely helpful with complex clients, checking in with your intention prior to a session (as a mindfulness therapist the default intention becomes "be in the body").
	Checking your intention as you report on a client to a colleague (e.g., asking "What is my intention in sharing this information?").
Attention	Could include Freud's suggestion for "evenly hovering attention".
	Being alert to the ways in which our attention to the client might be enhanced or diminished.
	Attending to our own mental and physical reactions during the interaction.
	Noticing any widening or narrowing of attention, moments when vividness becomes dullness.
Observation	Reminds us to keep asking questions and maintain our curiosity.
	Repeated sampling of the phenomena or interest (don't just take the first answer!).
	Listen and engage with "beginner's mind or the eyes of a child.
Compassion	Alert to moments when we can connect with our client's at the level of our common humanity—specifically how this unfolds in our body.
	Compassionate to ourselves in our work—acknowledging and accepting all the feelings our clients and colleagues evoke.
	Alert to and accepting of moments when compassion is lost.

References

Anderson, D. I., Campos, J. J., Witherington, D. C., Dahl, A., Rivera, M., He, M., Uchiyama, I., & Barbu-Roth, M. (2013). The role of locomotion in psychological development. *Frontiers in Psychology*, 4(440): 1–17.

App, B., McIntosh, D. N., Reed, C. L., & Hertenstein, M. J. (2011). Nonverbal channel use in communication of emotion: how may depend on why. *Emotion*, 11: 603–617.

Aron, A. R., Robbins, T. W., & Poldrack, R. A. (2004). Inhibition and the right inferior frontal cortex. *Trends in Cognitive Sciences*, 8(4): 170–177.

Azañón, E., & Haggard, P. (2009). Somatosensory processing and body representation. *Cortex: a Journal Devoted to the Study of the Nervous System and Behavior*, 45(9): 1078–1084.

Baer, R. A. (2005). *Mindfulness-Based Treatment Approaches: Clinician's Guide to Evidence Base and Applications*. Amsterdam, Boston: Academic Press.

Beck, J. S., & Beck, A. T. (2011). *Cognitive Behavior Therapy: Basics and Beyond*. New York: Guilford Press.

Berking, M., Wupperman, P., Reichardt, A., Pejic, T., Dippel, A., & Znoj, H. (2008). Emotion-regulation skills as a treatment target in psychotherapy. *Behaviour Research and Therapy*, 46(11): 1230–1237.

Berlucchi, G., & Aglioti, S. (1997). The body in the brain: neural bases of corporeal awareness. *Trends in Neurosciences*, 20(12): 560–564.

Bodhi, N. (2012). *The Numerical Discourses of the Buddha: a Complete Translation of the Auguttara Nikaya Teachings of the Buddha*. Somerville, MA: Wisdom.

Bohannon, J. (2010). Why do scientists dance? *Science*, 330(6005): 752–752.

Bresler, L. (2004). *Knowing Bodies, Moving Minds: Towards Embodied Teaching and Learning*. Dordrecht, Boston: Kluwer.

Brewer, J. A., Worhunsky, P. D., Gray, J. R., Tang, Y.-Y., Weber, J., & Kober, H. (2011). Meditation experience is associated with differences in default mode network activity and connectivity. *Proceedings of the National Academy of Sciences*, 108(50): 20254–20259.

Brookshire, G., & Casasanto, D. (2012). Motivation and motor control: hemispheric specialization for approach motivation reverses with handedness. *PLoS ONE*, 7(4): e36036.

Buffart, L. M., Uffelen, J. G. van, Riphagen, I. I., Brug, J., Mechelen, W. van, Brown, W. J., & Chinapaw, M. J. (2012). Physical and psychosocial benefits of yoga in cancer patients and survivors, a systematic review and meta-analysis of randomized controlled trials. *BMC Cancer*, 12(1): 559.

Cabral, P., Meyer, H. B., & Ames, D. (2011). Effectiveness of yoga therapy as a complementary treatment for major psychiatric disorders: a meta-analysis. *The Primary Care Companion to CNS Disorders, 13*(4). doi: 10.4088/PCC.10r01068.

Caldwell, K., Harrison, M., Adams, M., & Travis Triplett, N. (2009). Effect of Pilates and taiji quan training on self-efficacy, sleep quality, mood, and physical performance of college students. *Journal of Bodywork and Movement Therapies, 13*(2): 155–163.

Candidi, M., Aglioti, S. M., & Haggard, P. (2012). Embodying bodies and worlds. *Review of Philosophy and Psychology, 3*(1): 109–123.

Carlson, L. E. (2012). Mindfulness-based interventions for physical conditions: a narrative review evaluating levels of evidence. *ISRN Psychiatry, 2012*: 1–21.

Carmody, J., & Baer, R. A. (2008). Relationships between mindfulness practice and levels of mindfulness, medical and psychological symptoms and well-being in a mindfulness-based stress reduction program. *Journal of Behavioral Medicine, 31*(1): 23–33.

Chaddock, L., Hillman, C. H., Pontifex, M. B., Johnson, C. R., Raine, L. B., & Kramer, A. F. (2012). Childhood aerobic fitness predicts cognitive performance one year later. *Journal of Sports Sciences, 30*(5): 421–430.

Chapman, S. B., Aslan, S., Spence, J. S., DeFina, L. F., Keebler, M. W., Didehbani, N., & Lu, H. (2013). Shorter term aerobic exercise improves brain, cognition, and cardiovascular fitness in aging. *Frontiers in Aging Neuroscience, 5*. doi: 10.3389/fnagi.2013.00075

Chiesa, A., Serretti, A., & Jakobsen, J. C. (2013). Mindfulness: top-down or bottom-up emotion regulation strategy? *Clinical Psychology Review, 33*(1): 82–96.

Cicchetti, D., Ackerman, B. P., & Izard, C. E. (1995). Emotions and emotion regulation in developmental psychopathology. *Development and Psychopathology, 7*(1): 1–10.

Cloninger, C. R., Svrakic, D. M., & Przybeck, T. R. (1993). A psychobiological model of temperament and character. *Archives of General Psychiatry, 50*(12): 975–990.

Craig, A. D. (Bud). (2005). Forebrain emotional asymmetry: a neuro-anatomical basis? *Trends in Cognitive Sciences, 9*(12): 566–571.

Cummings, J. L. (1992). Depression and Parkinson's disease: a review. *The American Journal of Psychiatry, 149*(4): 443–454.

Damasio, A. R. (2012). *Self Comes to Mind: Constructing the Conscious Brain.* New York: Vintage.

Damasio, A. R., Grabowski, T. J., Bechara, A., Damasio, H., Ponto, L. L. B., Parvizi, J., & Hichwa, R. D. (2000). Subcortical and cortical brain activity during the feeling of self-generated emotions. *Nature Neuroscience, 3*(10): 1049–1056.

Davidson, R. J. (1992). Emotion and affective style: hemispheric substrates. *Psychological Science, 3*(1): 39–43.

Davis, C. M. (2009). *Complementary Therapies in Rehabilitation: Evidence for Efficacy in Therapy, Prevention,* and Wellness. Thorofare, NJ: SLACK.

Dehaene, S., & Naccache, L. (2001). Towards a cognitive neuroscience of consciousness: basic evidence and a workspace framework. *Cognition, 79*(1–2): 1–37.

Dillon, D. G., & Pizzagalli, D. A. (2007). Inhibition of action, thought, and emotion: a selective neurobiological review. Applied & Preventive Psychology: *Journal of the American Association of Applied and Preventive Psychology, 12*(3): 99–114.

Dusek, J. A., & Benson, H. (2009). Mind-body medicine: a model of the comparative clinical impact of the acute stress and relaxation responses. *Minnesota Medicine, 92*(5): 47–50.

Eberth, J., & Sedlmeier, P. (2012). The effects of mindfulness meditation: a meta-analysis. *Mindfulness, 3*(3): 174–189.

Ekman, P. (1992). An argument for basic emotions. *Cognition & Emotion, 6*(3–4): 169–200.

Farb, N. A. S., Segal, Z. V., Mayberg, H., Bean, J., McKeon, D., Fatima, Z., & Anderson, A. K. (2007). Attending to the present: mindfulness meditation reveals distinct neural modes of self-reference. *Social Cognitive and Affective Neuroscience, 2*(4): 313–322.

Fogel, A. (2013). *Body Sense: The Science and Practice of Embodied Self-awareness*. New York: W. W. Norton.

Fuchs, T., & Schlimme, J. E. (2009). Embodiment and psychopathology: a phenomenological perspective. *Current Opinion in Psychiatry, 22*(6): 570–575.

Gallagher, S. (2006). *How the Body Shapes the Mind* (new edn). Oxford: Oxford University Press.

Gard, T., Holzel, B. K., Sack, A. T., Hempel, H., Lazar, S. W., Vaitl, D., & Ott, U. (2012). Pain attenuation through mindfulness is associated with decreased cognitive control and increased sensory processing in the brain. *Cerebral Cortex, 22*(11): 2692–2702.

Gardini, S., Cloninger, C. R., & Venneri, A. (2009). Individual differences in personality traits reflect structural variance in specific brain regions. *Brain Research Bulletin, 79*(5): 265–270.

Germer, C. K. (2009). *The Mindful Path to Self-Compassion: Freeing Yourself from Destructive Thoughts and Emotions*. New York: Guilford Press.

Gibbs, R. W. (2006). *Embodiment and Cognitive Science*. Cambridge, New York: Cambridge University Press.

Gilbert, P., & Procter, S. (2006). Compassionate mind training for people with high shame and self-criticism: overview and pilot study of a group therapy approach. *Clinical Psychology & Psychotherapy*, 13(6): 353–379.

Gray, J. A., & McNaughton, N. (2000). *The Neuropsychology of Anxiety: an Enquiry Into the Functions of the Septo-hippocampal System*. Oxford: Oxford University Press.

Gross, J. J. (2001). Emotion regulation in adulthood: timing is everything. *Current Directions in Psychological Science*, 10(6): 214–219.

Haggard, P. (2005). Conscious intention and motor cognition. *Trends in Cognitive Sciences*, 9(6): 290–295.

Hammond, G. (2002). Correlates of human handedness in primary motor cortex: a review and hypothesis. *Neuroscience & Biobehavioral Reviews*, 26(3): 285–292.

Hayes, N. A., & Broadbent, D. E. (1988). Two modes of learning for interactive tasks. *Cognition*, 28(3): 249–276.

Hayes, S. C., Wilson, K. G., Gifford, E. V., Follette, V. M., & Strosahl, K. (1996). Experimental avoidance and behavioral disorders: a functional dimensional approach to diagnosis and treatment. *Journal of Consulting and Clinical Psychology*, 64(6): 1152–1168.

Hill, R., Ryan, P., Hardy, P., Anczewska, M., Kurek, A., Dawson, I., Laijarvi, H., Nielson, K., Nybourg, K., Rokku, I., & Turner, C. (2006). Situational levels of burnout among staff in six European inpatient and community mental health teams. *Journal of Mental Health Training, Education and Practice*, 1(1): 12–21.

Hillman, C. H., Erickson, K. I., & Kramer, A. F. (2008). Be smart, exercise your heart: exercise effects on brain and cognition. *Nature Reviews Neuroscience*, 9(1): 58–65.

Holzel, B. K., Lazar, S. W., Gard, T., Schuman-Olivier, Z., Vago, D. R., & Ott, U. (2011). How does mindfulness meditation work? Proposing mechanisms of action from a conceptual and neural perspective. *Perspectives on Psychological Science*, 6(6): 537–559.

Holzel, B. K., Ott, U., Gard, T., Hempel, H., Weygandt, M., Morgen, K., & Vaitl, D. (2007). Investigation of mindfulness meditation practitioners with voxel-based morphometry. *Social Cognitive and Affective Neuroscience*, 3(1): 55–61.

Hong, Y. (2008). *Tai Chi Chuan: State of the Art in International Research*. Basel: Karger.

Huberman, M., & Miles, M. B. (2002). *The Qualitative Researcher's Companion*. CA: SAGE.

Ives, J. C. (2003). Comments on "The Feldenkrais Method®: a dynamic approach to changing motor behavior". *Research Quarterly for Exercise and Sport*, 74(2): 116–123.

Jahnke, R., Larkey, L., Rogers, C., Etnier, J., & Lin, F. (2010). A comprehensive review of health benefits of Qigong and Tai Chi. *American Journal of Health Promotion: AJHP*, 24(6): e1–e25.

Kabat-Zinn, J. (1982). An outpatient program in behavioral medicine for chronic pain patients based on the practice of mindfulness meditation: theoretical considerations and preliminary results. *General Hospital Psychiatry*, 4(1): 33–47.

Kabat-Zinn, J. (2003). Mindfulness-based interventions in context: past, present, and future. Clinical Psychology: *Science and Practice*, 10(2): 144–156.

Kabat-Zinn, J., & Chapman-Waldrop, A. (1988). Compliance with an outpatient stress reduction program: rates and predictors of program completion. *Journal of Behavioral Medicine*, 11(4): 333–352.

Kashdan, T. B., Barrios, V., Forsyth, J. P., & Steger, M. F. (2006). Experiential avoidance as a generalized psychological vulnerability: comparisons with coping and emotion regulation strategies. *Behaviour Research and Therapy*, 44(9): 1301–1320.

Kerr, C. E., Sacchet, M. D., Lazar, S. W., Moore, C. I., & Jones, S. R. (2013). Mindfulness starts with the body: somatosensory attention and top-down modulation of cortical alpha rhythms in mindfulness meditation. Frontiers in *Human Neuroscience*, 7: 12.

Kerr, P. B., Caputy, A. J., & Horwitz, N. H. (2005). A history of cerebral localization. *Neurosurgical Focus*, 18(4): e1.

La Forge, R. (2005). Aligning mind and body: exploring the disciplines of mindful exercise. *ACSM's Health & Fitness Journal*, 9(5): 7–14.

Lan, C., Chen, S.-Y., Lai, J.-S., & Wong, A. M.-K. (2013). Tai Chi Chuan in medicine and health promotion. *Evidence-Based Complementary and Alternative Medicine: eCAM*, 2013: 502131.

Lazar, S. W., Kerr, C. E., Wasserman, R. H., Gray, J. R., Greve, D. N., Treadway, M. T., McGarvey, M., Quinn, B. T., Dusek, J. A., Benson, H., Rauch, S. L., Moore, C. I., & Fischl, B. (2005). Meditation experience is associated with increased cortical thickness. *Neuroreport*, 16(17): 1893–1897.

Leahy, R. L. (2006). *Contemporary Cognitive Therapy: Theory, Research, and Practice* (new edn). New York: Guilford Press.

Leahy, R. L., Tirch, D. D., & Napolitano, L. A. (2011). *Emotion Regulation in Psychotherapy: a Practitioner's Guide*. New York: Guilford Press.

LeDoux, J. E. (1996). *The Emotional Brain: the Mysterious Underpinnings of Emotional Life*. New York: Simon & Schuster.

Liao, C. M., & Masters, R. S. (2001). Analogy learning: a means to implicit motor learning. *Journal of Sports Sciences, 19*(5): 307–319.

Liao, W. (2000). *T'ai Chi Classics*. Boston, MA: Shambhala.

Liu, J., Li, B., & Shnider, R. (2010). Effects of tai chi training on improving physical function in patients with coronary heart diseases. *Journal of Exercise Science & Fitness, 8*(2): 78–84.

Longo, M. R., Azañón, E., & Haggard, P. (2010). More than skin deep: body representation beyond primary somatosensory cortex. *Neuropsychologia, 48*(3): 655–668.

Luders, E., Kurth, F., Mayer, E. A., Toga, A. W., Narr, K. L., & Gaser, C. (2012). The unique brain anatomy of meditation practitioners: alterations in cortical gyrification. *Frontiers in Human Neuroscience, 6*. doi: 10.3389/fnhum.2012.00034

Lutz, A., Brefczynski-Lewis, J., Johnstone, T., & Davidson, R. J. (2008). Regulation of the neural circuitry of emotion by compassion meditation: effects of meditative expertise. *PLoS ONE, 3*(3): e1897.

Lynch, D., Laws, K. R., & McKenna, P. J. (2010). Cognitive behavioural therapy for major psychiatric disorder: does it really work? A meta-analytical review of well-controlled trials. *Psychological Medicine, 40*(1): 9–24.

Mace, C. (2007). *Mindfulness and Mental Health, Theory and Science* (new edn). Hove, New York: Routledge.

Maiese, M. (2011). *Embodiment, Emotion, and Cognition*. Basingstoke, New York: Palgrave Macmillan.

McIver, S., O'Halloran, P., & McGartland, M. (2009). Yoga as a treatment for binge eating disorder: a preliminary study. *Complementary Therapies in Medicine, 17*(4): 196–202.

Mehling, W. E., Price, C., Daubenmier, J. J., Acree, M., Bartmess, E., & Stewart, A. (2012). The multidimensional assessment of interoceptive awareness (MAIA). *PLoS ONE, 7*(11)–e48230.

Mehling, W. E., Wrubel, J., Daubenmier, J. J., Price, C. J., Kerr, C. E., Silow, T., Gopisetty, V., & Stewart, A. L. (2011). Body awareness: a phenomenological inquiry into the common ground of mind-body therapies. *Philosophy, Ethics, and Humanities in Medicine, 6*(1): 6.

Meijer, M. (1989). The contribution of general features of body movement to the attribution of emotions. *Journal of Nonverbal Behaviour, 13*(4): 247–268.

Meltzoff, A. N. (1990). Towards a developmental cognitive science. *Annals of the New York Academy of Sciences*, 608(1): 1–37.

Meyer, T. D., & Hautzinger, M. (2012). Cognitive behaviour therapy and supportive therapy for bipolar disorders: relapse rates for treatment period and 2-year follow-up. *Psychological Medicine*, 42(7): 1429–1439.

Middleton, F. A., & Strick, P. L. (2002). Basal-ganglia "projections" to the pre-frontal cortex of the primate. *Cerebral Cortex*, 12(9): 926–935.

National Institute of Clinical Excellence (NICE) (2004). Depression: management of depression in primary and secondary care. Clinical guidelines, CG23 December 2004 (replaced by CG90).

Neff, K. D., & Germer, C. K. (2013). A pilot study and randomized controlled trial of the mindful self-compassion program. *Journal of Clinical Psychology*, 69(1): 28–44.

Niedenthal, P. M. (2007). Embodying emotion. *Science*, 316(5827): 1002–1005. doi:10.1126/science.1136930

Nietzsche, F. (1954). *Thus Spoke Zarathustra (A Thrifty Book): A Book for All and None*. Blacksburg: Thrifty Books, 2009.

Nithianantharajah, J., & Hannan, A. J. (2009). The neurobiology of brain and cognitive reserve: mental and physical activity as modulators of brain disorders. *Progress in Neurobiology*, 89(4): 369–382.

Noggle, C. A., & Dean, R. (2012). *The Neuropsychology of Psychopathology*. New York: Springer.

Pace, T., Negi, L., Donaldson-Lavelle, B., Silva, B. O., Reddy, S., Cole, S., Craighead, L., & Raison, C. (2012). P02.119. Cognitively-based compassion training reduces peripheral inflammation in adolescents in foster care with high rates of early life adversity. *BMC Complementary and Alternative Medicine*, 12(Suppl 1): P175.

Panksepp, J. (1998). *Affective Neuroscience: The Foundations of Human and Animal Emotions*. Oxford: Oxford University Press.

Piet, J., & Hougaard, E. (2011). The effect of mindfulness-based cognitive therapy for prevention of relapse in recurrent major depressive disorder: a systematic review and meta-analysis. *Clinical Psychology Review*, 31(6): 1032–1040.

Roth, A., & Fonagy, P. (2005). *What Works for Whom? A Critical Review of Psychotherapy Research*. New York: Guilford Press.

Russell, T. A. (2011). Body in mind training: mindful movement for severe and enduring mental illness. *British Journal of Wellbeing*, 2(4): 13–16.

Russell, T. A. (2015). *Body in Mind: Unlocking the Secrets of Mindfulness Through Movement*. London: Duncan-Baird/Watkins.

Ruths, F. (2011). Book review of *Mindfulness and Acceptance Based Behavioral Therapies in Practice* by Lizabeth Roemer and Susan M. Orsillo, NY:

Guilford Press, 2009. *Behavioural and Cognitive Psychotherapy*, *39*(1): 123–124.

Sandell, R., Blomberg, J., & Lazar, A. (1997). When reality doesn't fit the blueprint: doing research on psychoanalysis and long-term psychotherapy in a public health service program. *Psychotherapy Research*, *7*(4): 333–344.

Schore, A. N. (2011). The right brain implicit self lies at the core of psychoanalysis. *Psychoanalytic Dialogues*, *21*(1): 75–100.

Schwenkreis, P., El Tom, S., Ragert, P., Pleger, B., Tegenthoff, M., & Dinse, H. R. (2007). Assessment of sensorimotor cortical representation asymmetries and motor skills in violin players. *The European Journal of Neuroscience*, *26*(11): 3291–3302.

Segal, Z. V., Williams, J. M. G., & Teasdale, J. D. (2002). *Mindfulness-Based Cognitive Therapy for Depression: A New Approach to Preventing Relapse*. London: Guilford.

Shapiro, S. L., Brown, K. W., & Biegel, G. M. (2007). Teaching self-care to care-givers: effects of mindfulness-based stress reduction on the mental health of therapists in training. *Training and Education in Professional Psychology*, *1*(2): 105–115.

Sieh, R., & Ralston, P. (1993). *T'ai Chi Ch'uan: The Internal Tradition*. Berkeley, CA: Blue Snake.

Simeon, D., Guralnik, O., Hazlett, E. A., Spiegel-Cohen, J., Hollander, E., & Buchsbaum, M. S. (2000). Feeling unreal: a PET study of depersonalization disorder. *The American Journal of Psychiatry*, *157*(11): 1782–1788.

Tang, Y.-Y. (2011). Mechanism of integrative body-mind training. *Neuroscience Bulletin*, *27*(6): 383–388.

Tang, Y.-Y., & Posner, M. I. (2009). Attention training and attention state training. *Trends in Cognitive Sciences*, *13*(5): 222–227.

Tang, Y.-Y., Lu, Q., Fan, M., Yang, Y., & Posner, M. I. (2012). Mechanisms of white matter changes induced by meditation. *Proceedings of the National Academy of Sciences of the United States of America*, *109*(26): 10570–10574.

Thayer, J. F., & Lane, R. D. (2000). A model of neurovisceral integration in emotion regulation and dysregulation. *Journal of Affective Disorders*, *61*(3): 201–216.

Tomkins, S. S. (2008). *Affect Imagery Consciousness The Complete Edition*. New York: Springer.

Totton, N. (2005). *New Dimensions in Body Psychotherapy*. Maidenhead, New York: Open University Press.

Van Dam, N. T., Sheppard, S. C., Forsyth, J. P., & Earleywine, M. (2011). Self-compassion is a better predictor than mindfulness of symptom severity and quality of life in mixed anxiety and depression. *Journal of Anxiety Disorders*, 25(1): 123–130.

Vytal, K. E., Cornwell, B. R., Arkin, N. E., Letkiewicz, A. M., & Grillon, C. (2013). The complex interaction between anxiety and cognition: insight from spatial and verbal working memory. *Frontiers in Human Neuroscience*, 7: 93.

Wager, T. D., Lindquist, M., & Kaplan, L. (2007). Meta-analysis of functional neuroimaging data: current and future directions. *Social Cognitive and Affective Neuroscience*, 2(2): 150–158.

Wainapel, S. F., & Fast, A. (Eds.) (2003). *Alternative Medicine and Rehabilitation*. New York: Demos Medical.

Wallbott, H. G. (1998). Bodily expression of emotion. *European Journal of Social Psychology*, 28: 879–896.

Wang, C., Bannuru, R., Ramel, J., Kupelnick, B., Scott, T., & Schmid, C. H. (2010). Tai chi on psychological well-being: systematic review and meta-analysis. *BMC Complementary and Alternative Medicine*, 10(1): 23.

Wilkinson, J. (2013). *An Independent Audit of Evaluations of Body in Mind Training (BMT) for Health Professionals*. Health Academix Ltd.

Williams, J. M. G., & Penman, D. (2012). *Mindfulness: an Eight-week Plan for Finding Peace in a Frantic World*. Emmaus, PA: Rodale.

Wolpert, D. M., Diedrichsen, J., & Flanagan, J. R. (2011). Principles of sensorimotor learning. *Nature Reviews Neuroscience*, 12(12): 739–751. Available at: www.ted.com/talks/daniel_wolpert_the_real_reason_for_brains

Worsfold, K. E. (2009). The body in clinical cognitive theory: from Beck to mindfulness. *Contemporary Buddhism*, 10(2): 220–240.

CHAPTER EIGHT

Flow state experiences as a biopsychosocial guide for tai ji intervention and research in neuro-rehabilitation*

Giles Yeates

Introduction

This chapter will introduce the need for biopsychosocial interventions that can respond to multi-domain disturbances to self-states in neurological patients, states that have yet to benefit from existing psychological interventions. In particular tai ji (TJ) is suggested, with an emphasis on the subjective experience during practise. It is argued that the psychological dimension of TJ practise has been neglected in the existing literature on TJ with neurological conditions. The enhancement of flow state experiences (FSEs) is a goal within TJ practise, derived from the Daoist spiritual heritage of TJ. FSE is described (in its philosophical and practical aspects) and substantiated, linking to both parallel concepts in Western positive psychology and the existing research on TJ practise with neurological conditions. This chapter concludes by formulating suggestions for a future research programme to scientifically validate the underlying concept of FSE, in addition to targeting and evaluating its clinical application via TJ groups in neurological services.

---

* Originally published in 2015 in *Neuro-Disability & Psychotherapy*, 3(1): 22–41.

Fragmented vs. *coherent experiences in neurological conditions*

If something is broken into many pieces, how can wholeness and coherence be achieved once more? One possibility is when those pieces become assimilated into a higher-level process or form, where they are bound together through the forces or structures inherent in the higher-level process. Fragments of paper being held in the same trajectory of movement or shape, within the current of air or water would be a good example.

Fragmented, oscillating, and incoherent self-states characterise many neurological conditions (e.g., Charmaz, 1990; Luria, 1975). Other experiences involve inertia, and a mismatch of intention and action without a cue or external stimulus (e.g., Sacks, 1973). In addition to the aims of western talking psychotherapies with neurological groups, a newer focus on shifting the relationship of a person to their mental contents has been introduced by psychological interventions influenced, in part or whole, by Eastern spiritual traditions (e.g., Bedard et al., 2012; Deepak et al., 1994; Lynton et al., 2007; Yeates & Farrell, 2014). Targeting ruminative content-filled minds in neurological conditions, stillness is often strived for in response to busy, cluttered psychological experiences.

However, fragmentation and inertia in self-experience have not been targeted by this class of intervention to date. Furthermore, it is possible that some of these very issues have challenged many service users' ability to tolerate and benefit from aspects of the mother Eastern traditions. For example, the requirements of embodied stillness and concentration in sitting meditation practices may exclude many (Russell, 2011; Russell & Tatton-Ramos, 2014). Many people with neurological conditions are predominantly distressed/preoccupied by their embodied and mobility experiences (e.g., regaining functional limb use following hemiparesis in stroke; initiating or terminating movement in Parkinson's disease; managing balance difficulties across a range of conditions), and would not find face validity or relevance in a talking or meditative therapy.

Flow and change practices in Daoist traditions

Mindfulness and non-judgmental relationship to mental events and desires, concepts and practices, derived from Buddhist and Vedic ideas (Kabat-Zinn, 2003), have been applied widely within western psychotherapy and health settings, including neurological services (Bedard et al., 2003, 2005, 2012; Detert & Douglas, 2014; Grossman et al., 2010; Johansson et al., 2012). While other major Eastern theological and philosophical systems may offer similar applied potential, they have yet to be exploited. Gains reported for Vedic yoga practices in neuro-rehabilitation (Bastille & Gill-Body, 2004; Lundgren et al., 2008; Lynton et al., 2007; Rajesh et al., 2006; Shravat, 2014; Yeates et al., 2015) suggest that embodied mindfulness approaches beyond sitting meditation may be of value, as suggested by Russell (2011; Russell & Tatton-Ramos, 2014).

A further heritage to explore for future clinical application is Daoism (Taoism), practised by at least twenty million people in China and worldwide. An early and long-lasting system of thought (formalised in the second century CE), Daosim is thought to have developed from prehistoric shamanic beliefs. Its core features are communicated in various bodies of literature, famously including the Dao De Ching, attributed in Chinese folklore to a semi-historical/ mythological figure, Lao Tzu (Kaltenmark, 1969). In ancient Daoist writings, philosophical ideas were communicated powerfully through imagery and allegory, famously including the Zhuangzi's (fourth century BC) butterfly dream:

> Once upon a time, I, Chuang Chou, dreamt I was a butterfly, flutter-ing hither and thither, to all intents and purposes a butterfly. I was conscious only of my happiness as a butterfly, unaware that I was Chou. Soon I awoke, and there I was, veritably myself again. Now I do not know whether I was then a man dreaming I was a butterfly, or whether I am now a butterfly, dreaming I am a man. Between a man and a butterfly there is necessarily a distinction. The transition is called the transformation of material things. (cited in Merton, 1969)

This account conveys a central Daoist concept: the impermanence of states in the universe, and the constant transition from one state/form to another as part of the endless flow of universal forces. This was represented by Zhou Dunyi (1017–1073) in original diagrammatic form, the taijitu ("the diagram of the supreme ultimate"). An example is presented in Figure 1.

The presence of a black circle inside the white section and vice-versa represents motion and the constant state of change in the harmonising

Figure 1: Tajitu diagram.

and re-balancing between two different types of universal forces, form/presenting/pushing/pervading/punctuating (Yang) and yielding/embracing/incorporating/receiving (Yin). The surrounding eight trigrams represent constituent universal eight dimensions/expressions organised within these two meta-principles. Those pursuing a Daoist spiritual framework, such as the priests, monks, and nuns in isolated temple complexes around China, aim to cultivate internal health and well-being through Neidan, "inner alchemy" processes and the harmonisation of their life patterns and activities within the flow of wider forces in the universe. This internal cultivation is conceptualised within a Daoist framework as a continual process of converting lower forms of energy (taken from the external environment via breath and receptivity to external sources) into progressively higher ones (qi to jing to shen) at key bodily locations, and finally returning the reified energy form back into the external universe, so completing a cycle.

One aspired principle and subjective experience within this harmonisation is called Wu Wei: the practice of non-action, an uncontrived non-effortful experience of oneself to be in harmony with greater universal forces. Methods to achieve and maintain this harmonisation include meditation, diet, music, horticulture, architecture, communing with nature, and exercise/movement regimens. A key quality of this simultaneous view of the universe and body (as an element of the universe) is the priority of flow, the movement and transition of one state into another. In martial arts and wider philosophies, Bruce Lee (explicitly acknowledging a Daoist heritage) has offered famous injunctions to prioritise flow. These include:

> Be like water . . . empty your mind, be formless, shapeless like water. If you put water in the cup, it becomes the cup. You put water in the bottle, it becomes the bottle. You put it in the teapot, it becomes the teapot. Now, water can flow or it can crash. Be water my friend. (Little & Lee, 2000, TV documentary)

> We are always in a process of becoming and nothing is fixed. Have no rigid system in you, and you'll be flexible to change with the ever changing. Open yourself and flow, my friend. Flow in the total openness of the living moment. If nothing within you stays rigid, outward things will disclose themselves. Moving, be like water. Still, be like a mirror. Respond like an echo. (Lee, 2000, p. 13).

Outside of traditional Chinese medicine (and the Western evaluation of its specific treatments), the applied benefits of underlying Daoist concepts and practices have received relatively little attention.

The psychological study of flow states

While the validity and application of the esoteric energetic theory inherent in Daoist health cultivation will not be considered here, the subjective dimension of harmonised flow in Daoist mind-body practices is a key focus. This overlaps significantly with the concept of flow states conceptualised in Western positive psychology, in the work of Mihály Csíkszentmihályi (1990, 1997) in particular. He has studied intense states of absorption in the domains of sport, music, creativity, and work. Flow states are routinely described by practitioners as involving the dissolving of a self-state and loss of normal self-boundaries (loss of reflective self-consciousness), distortion of temporal experience, a merging of action and awareness where intention is not effortful and the activity concerned seems to flow forth of its own accord. Practitioners feel intense well-being, ecstatic experiences at the time, and part of something bigger than themselves. Importantly, those experiencing flow states in a particular activity have attained some level of mastery over that activity through practise and experience, such that there is a dimension of automaticity and diminution of effort. Practitioners across diverse fields consistently use a metaphorical language of creativity and action flowing forth, hence Csíkszentmihályi's term flow states.

Csíkszentmihályi acknowledged historical precursors to his psychological study of flow, notably religious thought from near and far eastern traditions. Many faiths have historically articulated and practiced a desired goal to be at oneness with something greater, via dissolution of the boundaries of present sensory experience. In addition to Daoism these include: in Islam, Sufi Dervishes' practice of spinning and whirling; in Vedic/Hindu thought a form of flow-like psychological absorption in an object of meditation, Samyama, aimed for in Raja Yoga and in fast forms of physical yoga flow states are explicit as the desired vehicle for self-transcendence; Gregorian chanting in Christianity and the use of prayer beads and repetition of prayer forms/mantras in all the major faiths.

While there is clearly overlap between Daoist and positive psychological conceptualisation of FSEs, the latter differs in emphasising explicit appraisal as a critical dimension. This includes assertions of flow components and conditions for flow such as an individual's sense of personal control or agency, involvement in a task with a clear set of goals and progress, and which provides immediate feedback, simultaneous perception of high task challenges alongside high personal skill (Csíkszentmihályi et al., 2005; Shaffer, 2013). While Daoist masters clearly have mastery of highly skilled practices such as martial arts, accounts of their subjective experiences during practise indicate a diminution of explicit appraisal processes generally and subjective absence of perceived challenge or personal skill (see The Taijiquan Classics, Davis, 2004).

Tai ji quan

Tai ji quan (TJ) means "supreme ultimate fist" (often spelled tai chi in the English language), an expression through movement of the principles of harmony and change in the Daoist view of the universe. It is characterised by soft flowing movements (done from either standing or sitting positions), punctuated by sudden explosive movements in some styles. These movements are synchronised with enhanced proprioceptive and interoceptive awareness, breath control, and attentional focus. The practitioner aims to still their mind through movement, an immersion into a broader form (the shape and sequence of movements) and force (the flow, direction, and emphases of the movements). The shape and sequence of movements become over-learned through many repetitions to allow progressive automatisation, removal of conscious interference, and the enhancement of a subjective state absorbed in the present moment, in the flow of movements. There is a simultaneous holding of "inner stillness" alongside movement, which rather than a contradiction is described in a classical Daoist text as a "still-point" or "pivot" in the flow changing states (Daoshu):

> So is there really a "this" and "that"? Or is there really no "this" and no "that"? When there is no more separation of "this" and "that", we have what is called the still-point of the Dao. (Zhuangzi Qiwulun, cited in Merton, 1969)

Within the mind-body practice of TJ, the subjective quality of flow is associated with embodied feelings of lightness and bodily coherence/ unity within movement, as indicated in these excerpts from the earliest attributed writing on TJ:

> In motion the whole body should be light . . . with all parts of the body linked as if threaded together . . . All movements are pivoted by I (mind-intention), not external form . . . If the I wants to move upward, it must simultaneously have intent downward . . . the whole body should be threaded together through every joint without the slightest break . . . like a great river rolling on unceasingly. (Zhang San Feng, 1279–1386, cited in Davis, 2004)

The "I" in these writings should not be confused with an over-thinking prioritised self. While "intention" is a critical aspect of TJ practice, this is an "in-the-moment" distributed focus, not an a priori top-down anticipatory process. Increased awareness of internal and external stimuli is recruited in the optimisation of the practitioner's lightness and effortlessness (*Wu Wei*) of movement.

One common tradition holds that TJ was developed in the twelfth or thirteenth century as a martial art in Wudang Mountains, Hubei (Central China) by the author of the last quote, Daoist monk Zhang San Feng. He had a revelation when seeing a snake fight a bird (most commonly narrated as a crane) and effectively defending itself through yielding to and absorbing the attacking moves of its aerial attacker. Over nearly a millennium TJ has evolved into many different types of schools and styles across China, named after the families that have developed them. These vary significantly in emphasis, force, speed, and postures, and include the more widely known chen, yang, and wu styles, in addition to that still practised by the monastic communities on Wudang mountains and those schools that claim direct ancestry from them.

Regular practitioners have historically learned TJ for both martial skills and simultaneous health benefits. TJ and closely related practices known as chi/qi gong are considered within the hydraulic metaphors of traditional Chinese medicine theory to free the flow and circulation of vital life force and energy (chi) from blockages in the body, thereby balancing health. Scientific studies of TJ practitioners in the general population do reliably highlight gains in physical (muscle strength, balance, flexibility, and motor control) and cognitive functioning (executive functioning) plus emotional well-being (see Jahnke et al., 2010; Wei et al., 2013, for review). In addition, long-term TJ practitioners are characterised by particular physiological structural differences, including bone density (Jahnke et al., 2010) and increased tissue volume in specific cortical areas (Wei et al., 2013, 2014). Wei and colleagues found that when compared to matched controls, long-term TJ practitioners (average 14 ± 8 years) were shown to have increased volume in precentral gyrus, insula sulcus, and middle frontal sulcus in the right hemisphere and superior temporal gyrus, medial occipito-temporal sulcus, and lingual sulcus in the left hemisphere. Furthermore, the TJ group had increased functional homogeneity (improved functional integration) in the right precentral gyrus and decreased functional homogeneity (improved functional specialisation) in the left anterior cingulate. The tissue volume of the left medial occipito-temporal sulcus and lingual sulcus was positively correlated with the intensity of TJ practise (hours per week). Collectively, this data suggest unique gains and processes inherent to TJ practise that are distinct from known benefits from sitting meditation practices.

Other studies have investigated the efficacy of TJ in improving aspects of functioning for various clinical groups. A review of studies across a range of long term health conditions highlighted increased cardiovascular fitness, balance control, and flexibility (Wang et al., 2004). Alongside these gains in physical functioning, the emotional benefit of TJ has long been acknowledged. Reduced stress, anxiety, depression, mood disturbance, and increased self-esteem of practitioners have been reported by Wang and colleagues (2010).

Tai ji quan in neurological services

When TJ is considered as a historical practice of FSE cultivation, the evaluation of TJ in neurological services may indicate the clinical relevance of such states. Clinical gains from TJ group classes are indeed reported within the neurological literature: group studies of survivors of traumatic brain injury using a control condition have produced favourable results (Blake & Batson, 2009; Gemmell & Leathem, 2006), building on earlier single case observations by Shapira and colleagues (2001). Gemmell and Leathem (2006) noted that survivors of traumatic brain injury reported feeling less tense, angry, afraid, confused, sad, and more energetic and happier following the sessions. Blake and Batson (2009) demonstrated gains for a TJ group relative to exercise-only controls in the areas of mood and self-esteem, but not physical functioning. In stroke research, TJ interventions have been shown to lead to mood and general functioning gains alongside no adverse effects for community-based participants (Hart et al., 2004), improve standing balance in elderly stroke patients (Au-Yeung et al., 2009), and a review article and pilot study recommended TJ to be a core component of stroke services (Taylor-Piliae & Coull, 2012; Taylor-Piliae & Haskell, 2007).

Concurrent physical and psychological gains are also reported in progressive neurological conditions. A pilot study focusing on multiple sclerosis (MS) indicated improvements on measures of balance and de-pression following a TJ/qi gong intervention (Mills et al., 2000). A study of a twice-weekly six month TJ intervention for MS reported indications of gains in balance, coordination, and depression, relative to a treatment-as-usual group (Burschka et al., 2014). In two other TJ MS studies, gains in walking distance, hamstring flexibility, and psychological well-being were reported by Husted and colleagues

(1999), while improved perceptions of physical and mental health (but not mobility) were reported by Tavee and colleagues (2011). Participants with Parkinson's disease demonstrated gains in postural stability, mobility, functional reaching, and balance in single case, pilot, and randomised group studies (Hackney & Earhart, 2008; Li, 2013; Li et al., 2007; 2012; Venglar, 2005) alongside answering favourably on a nonstandard questionnaire of mood and well-being (Hackney & Earhart, 2008). Quinn and Jones (2012) report psychological and physiological gains across participants in a heterogeneous neurological disability sample.

While this evidence is collectively encouraging, recent critiques of the scientific TJ literature as a whole have drawn attention to a range of methodological weaknesses. These include small sample sizes, poorly described treatment protocols, and the use of inadequate outcome measures (Zhang et al., 2011). Furthermore each research-clinical team across studies has selected different movement sequences from varied TJ styles to build their intervention protocol, making comparison across studies and generalisation of results problematic. In addition, and significantly for the focus of this chapter, many of the aforementioned studies have favoured physical functioning as a target for TJ intervention and outcome evaluation. Outcomes in emotional functioning have often remained absent or have been operationalised without a coherent theoretical framework linking these to other aspects of functioning, nor the interrelationship of TJ to emotional experience and other facets of self-experience during TJ practise.

Burschka and colleagues (2014) have offered the most elaborated theoretical account of the subjective dimension of TJ practice and its importance to neurological conditions. These authors make specific associations between sustained and switching attentional processes, acceptance, non-elaborated awareness focus during embodied meditative practice, and gains for cognitive and physical functioning, fatigue, depression, anxiety, and well-being. They have based this model on a parallel account of mindfulness meditation (Bishop et al., 2004) and have categorised TJ as a form of mindfulness intervention, operationalising similar key processes. As noted above mindfulness has become a familiar concept in Western health care settings, and overlaps with many aspects of TJ practice (non-judgmental acceptance of experience and non-elaborated awareness in an over-thinking cognitive sense are crucial dimensions of TJ), so has face validity as a candidate for an internal psychological dimension of TJ practice.

A greater awareness of the Daoist spiritual context of TJ would caution against a simple application however. While some of the *neidan* practices are closer to Buddhist mindfulness meditation, TJ practice itself invites particular subjective characteristics for the practitioner (see above), and the Daoist focus on stillness and coherence within transitory, changing states that pervades all mind-body practices seems to be qualitatively different than mindfulness' non-judgmental acceptance of the present moment. While also a route away from self-judgmental experience, Daoist TJ attaches value to an aspired experience of coherence, inter-connection with something bigger, and flow as part of transition (but similarly discourages in-the-moment self-criticism in practitioners who feel they are falling short of such goals). This focus offers a distinct rationale for mind–body practices in response to distressing experiences of fragmented self-states and inertia in neurological conditions. However, the aforementioned Daoist meta-physical frameworks, primers for specific subjective experience during practise and the dimension of spirituality as a whole is notable in its absence from the TJ literature. A UK TJ neuro-rehabilitation research group have developed an unpublished holistic evaluation measure of TJ practise, the Tai Chi Movement for Wellbeing Effectiveness Measure (Quinn, unpublished), covering physical and psychological functioning. This approach to TJ evaluation remains in the minority, however.

In summary, when outcomes have been sufficiently broadened, there is encouraging early evidence from existing controlled group studies to support TJ as a biopsychological intervention, with reported gains across physical and psychological domains in a number of neurological patient groups. However, further work is needed to sensitively and reliably capture psychological and spiritual dimensions within TJ research, in addition to social dimensions (e.g., group identity/membership/affiliation experience) to fully document the biopsychosocial value of TJ. With these outstanding areas of development in mind, the study of FSEs in TJ, drawing on the Daoist heritage of this practice, may be well positioned to fill these gaps and also connect with the wider contemporary interest in the clinical value of meditative states, while retaining a distinct contribution among other traditions.

Heterogeneity in clients' needs as a challenged to tai ji practice

As noted by Yeates and colleagues (2015) in relation to Yoga interventions within neurological services, within many acquired and progressive neurological categories, there is considerable diversity in each client's constellation of physical, cognitive, and emotional needs (Yeates & Farrell, 2014). These can uniquely challenge the standardised implementation of a body–mind intervention. For example, many clients will need to learn chair-based sequences due to mobility and/or balance difficulties. Those with fatigue may need shorter sessions. Someone with dyspraxia may not be able to match a visual demonstration of a posture/sequence to their own intentional movement, and as such would need a hands-on adjustment from the teacher to learn through proprioceptive feedback. Another person's memory difficulties may mean that learning a long sequence may best be achieved through the use of visual/written prompts and/or learning the sequence in reverse order via backward chaining or other errorless learning approaches (Wilson et al., 2010).

If anxiogenic rumination is a potential barrier to a mindful present-focus during TJ practise, then the style of teaching needs to be adapted accordingly, minimising explicit knowledge transfer and emphasising a greater degree of proprioceptive embodied focus. To return to the self-state disturbances discussed in the opening section, over-learning and repetition can be optimised for those with initiation difficulties, and TJ sequences characterised by regularities selected for those whose distress follows common incoherence and fragmentation in self-experience.

Given this complexity in learning and adaptation considerations, it is striking that there is very little mention of such issues in the TJ neurological literature, where often the group under study is directly or indirectly conveyed as homogenous. Furthermore, with the exception of Burschka and colleagues (2014), most interventions studied are often around ten to fifteen weekly sessions and so are unlikely to allow sufficient time for either adequate bespoke learning of movements across heterogeneous participants' needs, nor the progressive automaticisation of movement and attainment of FSE.

This diversity in participants' needs is both a challenge to the generalisability of TJ research and an essential clinical consideration in identifying TJ sequences that will be accessible to each client. With reference to the positive psychology literature on flow, it is easy to see how an individual trying to learn a sequence that is perceived as too difficult to master, which may lead to a self-critical inner dialogue during practise, will prevent the progressive attainment of FSE. Similarly if a sequence lacks meaning, experienced as unchallenging or irrelevant, it will lack a quality of attentional capture and absorption and equally will not be a conduit of FSE emergence. As noted above, Russell (2011; Russell & Tatton-Ramos, 2014) claims that embodied mindfulness practices such as TJ offer greater accessibility to wider groups than the attentional demands in forms of sitting meditation. This may depend on the mediation of person-sequence fit and FSE attainment.

Positive psychological study of flow articulates the need for balance in each practitioner's perception of task demands *vs.* perceived personal skills as a precondition for FSE emergence. As the concept of FSE articulates the need for this balance in this precise way, this is in fact suited as a guiding principle for multiple professions working across physical, cognitive, and emotional domains to inform the bespoke selection of TJ sequences for each client.

Characteristics of a flow-orientated tai ji group for neurological conditions

A FSE-orientated TJ intervention for people with neurological conditions, when founded on the above principles would have the following elements:

- It would be a group intervention. Aside from economic advantages, the experience of social fellowship, affiliation, and at times synchronised movement with others are all in keeping with FSE optimisation.
- This is envisaged as a long-term community intervention (minimum six months to a lifetime), to allow increasing familiarity and the progressive automaticisation of movements to occur over a substantial time period. The increasing recognition of long-term community exercise and well-being groups for neurological groups, such as the recent proliferation of ARNI (action for rehabilitation from neurological injury) groups for stroke survivors in

the UK (Balchin, 2011; www.arni.uk.com) would be consistent with a FSE TJ remit.

- A typical session will contain both individual work and group synchronised movements. The individual work section of the session will allow the opportunity for a participant to work on their own TJ sequence. This would be chosen for optimal personal ability-task requirement fit and adapted in response to each participant's unique physical, cognitive, and emotional needs. This section would be sandwiched in between group warm up sequences (chosen for universal accessibility) and a synchronised group TJ sequence at the end, again chosen for universal accessibility. The experience of being within a shared group sequence is a further opportunity for FSE attainment and transcendence beyond individual experience.

- Typical of most TJ classes, the sessions would be at least weekly and last sixty minutes. All participants would be encouraged to do the warm-up and end group sequence. However the middle individual work section will vary across individuals in the ratio of activity/rest time, as a function of individual fatigue/pain/ physical management needs.

- Where local weather and locations' disability access allow, opportunities should be sought to teach and practice outside in a natural setting, to allow practitioners to use their sensory experience of natural stimuli (visual, auditory, tactile sensory awareness, e.g., the sound of water moving, the sensation and sound of a breeze in the trees) as experiential drivers to attain FSE (i.e., an experience of being at one with natural processes bigger than oneself).

- Participants would be encouraged to develop a concurrent self-practise at home to increase repetitions of the sequences and progressive automatisation of the movements conducive to FSE attainment. This personal practise would be supported by multimedia materials to facilitate learning.

- The constituent TJ sequences would be drawn and adapted from classic TJ styles and forms, as per the training/experience of the instructor/facilitator. However, participants would be additionally encouraged to maximise personal experience of flow and absorption in the movements, while minimising self-consciousness and any self-critical inner dialogue.

Hypothesised value of FSEs through TJ practise in neurological services

Seven broad areas of clinical value are hypothesised to follow TJ provision across neurological groups, aimed at maximising FSEs. The initial points share a focus on immediate client gain while the latter two offer concurrent gains for research evaluation and service organisation:

1. Subjective FSEs, increasingly accessed through the repetition of appropriate TJ movements, will be pleasurable and comforting to people with neurological conditions who experience common disturbances to self-states (such as fragmentation, uncontrollable rumination, and inertia). The increasingly predictable attainment of FSEs will be associated with reductions in common forms of psychological distress such as depression, anxiety and anger.

2. The movement and bodily aspect of TJ will offer greater attentional capture and promotion of engagement for many people than a sitting meditation practise (Russell, 2011; Russell & Tatton-Ramos, 2014).

3. Regardless of an individual's focus on physical or psychological changes as part of their neurological condition, FSE-orientated TJ will offer face-validity as a relevant intervention that will actually lead to gains in both areas, avoiding any Cartesian dichotomous tension between body *vs.* mind focus. Reports across studies of superior performances on executive functioning tests in TJ practitioners within the general population suggest at least a potential relevance of the practice for cognitive functioning in neurological groups too.

4. The selection of TJ postures, sequences, and styles in both clinical practice and research evaluation could be led by FSE attainment as the critical driver. The positive psychology FSE components of matched individual ability and task demands can be used to calibrate and reduce the discrepancy between the two variables during the bespoke selection and learning of TJ sequences across individuals. Different individuals may be working on different sequences but aiming towards the same relative FSE attainment, progressing in this as a function of practise. Movement from physically easier to more demanding sequences

would also be paced by FSE attainment and maintenance in each sequence.

5.   The bespoke matching of postures and sequences to a client's biopsychosocial needs in order to optimise FSE attainment is truly an interdisciplinary team (IDT) process, with physiotherapists considering the physical demands of sequences in relation to client ability; psychologists similarly considering the cognitive and emotional demands; occupational therapists evaluating participation and activity meaning; and in relation to some clients all team members holding in mind the spiritual significance of FSE attainment, depending on the client's belief systems.

6.   As such FSE attainment and maintenance is very usefully placed as a shared IDT goal for a neurological services team and may offset common service trends to further fragment a client's experience in an iatrogenic way, by partitioning physical, cognitive, emotional, and social needs, alongside a frequent neglect of spirituality (Yeates et al., 2015). This IDT service approach is supported by TJ training models for clinicians already operating in public health services (e.g., www.balancedapproach.co.uk/rehab/tai-chi-for-rehabilitation. html www.tmwtraining.com).

7   FSE attainment will therefore be a stable point of comparison across TJ intervention studies or clinical audits, regardless of TJ style or sample characteristics.

Critical steps in a future tai ji flow state experiences research programme

First, the attainment of FSEs needs to be reliably described, to guide interventions and validate outcome and process measurement. This would require both qualitative investigations of subjective experience during FSE in TJ practise, in addition to objective measurement. The use of a validated questionnaire of flow states from positive psychology, the Flow Conditions Questionnaire (FCQ, Schaffer, 2013), may be additionally valuable for TJ research, but given the aforementioned lack of absolute fit between positive psychological and Daoist concepts, some adaptation of questionnaire items and revalidation (including tailoring to specific clinical groups) may be required.

Another approach would be to use psychophysiological measurement to reliably characterise FSE attainment, following existing research that has identified characteristic alpha and theta brain rhythms of experienced meditators in Himalayan Buddhist communities (e.g., Lyubimov, 1998). The use of mobile electroencephalographs (EEGs) to study neural patterns in Daoist priest TJ masters performing forms and sham movement sequences, in comparison to intermediate and novice TJ practitioners would be an interesting way of elucidating the psychophysiological characteristics of FSEs. Particular cardiac activity signatures during trained meditation states (Kim et al., 2013) and in other states of well-being (McCraty et al., 1995) have been reported in the psychophysiological literature. Kim and colleagues (2013) have observed increased concordance of alpha brain rhythms and heart-rate coherence (reduced variability in time intervals between heart beats) during meditation. This simultaneous real-time neural and cardiac activity measurement using multi-channel recorders would offer further ways to objectively characterise FSEs during TJ practise.

With FSE subjective and psychophysiological characteristics reliably described, the interrelationship of FSE attainment and neuro-disability may also be richly explored using mixed methods. Qualitative studies of long-term TJ practitioners with both chronic and recently acquired disabilities may aid the conceptual unpicking of this relationship. Correlational studies of a validated FSE questionnaire data in each major neurological group and associations with measures of physical, cognitive, emotional, and spiritual functioning would be of interest. A detailed observational study of a standard unmodified TJ class for differing neurological groups would substantiate some of the claims made above. Instances of the learning and maintenance of TJ being challenged by different kinds/combinations of neurological difficulties (physical, cognitive, emotional) could be coded and tallied. The relationship of this data to FSE attainment as measured by repeated sessional questionnaires could be explored. In addition any spontaneous adaptation strategies used by a TJ instructor in response to these difficulties could be recorded and evaluated.

Finally, TJ interventions for neurological groups based on the IDT bespoke matching of sequences to individuals' unique constellation of biopsychosocial needs to optimise progressive FSE attainment would require formal evaluation. This could be via a randomised trial comparison with standardised but unadapted TJ provision and a control group exercise condition, focusing on a range of biopsychosocial outcomes of interest. These would include pre-post and repeated sessional questionnaire measures of anxiety, depression, anger, fatigue, flow, and well-being. Following Burschka and colleagues (2014) group interventions of six months minimum are recommended as a parameter of evaluative studies. Briefer evaluations run the risk of reflecting insufficient time spent on bespoke learning and automaticisation of the movements to access FSE. This longer time focus will also be consistent with the remits of many community service interventions. Acceptance of these group formats and the FSE-focus for neurological patients would be measured via recruitment and drop-out rates, and participant feedback. Pre-post intervention changes in physical and cognitive functioning would be a further comparison across interventions, and of interest given gains reported in the generic scientific TJ literature.

Conclusion

A case has been made for the clinical value of concept of flow state experience attainment during tai ji practise, with neurological groups in particular. FSEs have been prioritised in both Daoist spiritual and positive psychology literature, with both overlap and differences between these. It is suggested that FSEs uniquely respond to subjective experiences of fragmentation and incoherence in many neurological conditions, and that the properties of TJ forms may support these and other self-state disturbances, such as inertia. The existing neurological literature supports the clinical value of TJ in producing gains to physical and emotional functioning across a range of conditions. However, an underlying biopsychosocial theoretical framework linking these dimensions of outcome is absent and spiritual aspects neglected. FSEs are advocated as a valuable nodal point to link these dimensions and also practically guide the bespoke adaptation of TJ sequences to varied physical, cognitive and emotional difficulties within neurological groups. Finally, a programme to scientifically validate and apply these concepts clinically has been formulated.

Acknowledgements

I would like to thank Dr David Quinn for his time, comments, and hospitality in increasing my understanding of how TJ can play a pivotal role in an NHS neuro-rehabilitation service. I also offer respect to my Shifu, Zhong XueChao, Daoist Priest and fifteenth generation Wudang Taiji and kung fu master at Five Dragons Temple, Wudang Mountains. His historical inspiration, tuition, and encouragement have been the life-force behind these ideas.

References

Au-Yeung, S. S. Y., Hui-Chan, C. W. Y., & Tang, J. C. S. (2009). Short-form Tai Chi improves standing balance of people with chronic stroke. *Neurorehabilitation & Neural Repair, 23*: 515–522.

Balchin, T. (2011). *The Successful Stroke Survivor*. Lingfield: Bagwyn.

Bastille, J. V., & Gill-Body, K. M. (2004). A yoga-based exercise program for people with post-stroke chronic hemiparesis. *Physical Therapy, 84*: 33–48.

Bedard, M., Felteau, M., Gibbons, C., Klein, R., Mazmanian, D.. Fedyk, K., & Mack, G. (2005). A mindfulness-based intervention to improve quality of life among individuals who sustained traumatic brain injuries: one-year follow-up. *The Journal of Cognitive Rehabilitation, Spring*: 8–13.

Bedard, M., Felteau, M., Marshall, S., Dubois, S., Gibbons, C., Klein, R., & Weaver, B. (2012). Mindfulness-based cognitive therapy: benefits in reducing depression following a traumatic brain injury. *Advances in Mind–Body Medicine, 26*: 14–20.

Bedard, M., Felteau, M., Mazmanian, D., Fedyk, K., Klein, R., Richardson, J., Parkinson, W., & Minthorn-Biggs, M. B. (2003). Pilot evaluation of a mindfulness-based intervention to improve quality of life among individuals who sustained traumatic brain injuries. *Disability and Rehabilitation, 25*: 722–731.

Bishop, S. R., Shapiro, S., Carlson, L., Anderson, N. D., Carmody, J., Segal, Z. V., Abbey, S., Speca, M., Velting, D., & Devins, G. (2004). Mindfulness: a proposed operational definition. *Clinical Psychology: Science & Practice, 11*: 230–241.

Blake, H., & Batson, M. (2009). Exercise intervention in brain injury: a pilot randomized study of Tai Chi Qigong. *Clinical Rehabilitation, 23*(7): 589–598.

Burschka, J. M., Keune, P. M., Oy, U., Oschman, P., & Kuhn, P. (2014). Mindfulness-based interventions in multiple sclerosis: beneficial effects of Tai Chi on balance, coordination, fatigue and depression. *BioMedCentral Neurology, 14*(1): 165.

Charmaz, K. (1990). "Discovering" chronic illness: using grounded theory. *Social Science & Medicine, 30*: 1161–1172.

Csíkszentmihályi, M. (1990). *Flow: The Psychology of Optimal Experience.* New York: Harper & Row.

Csíkszentmihályi, M. (1997). *Finding Flow: The Psychology of Engagement With Everyday Life.* New York: Basic Books.

Csíkszentmihályi, M., Abuhamdeh, S., & Nakamura, J. (2005). Flow. In: A. Elliot (Ed.), *Handbook of Competence and Motivation* (pp. 598–698). New York: Guilford Press.

Davis, B. (2004). *The Taijiquan Classics: An Annotated Translation.* Berkeley: North Atlantic.

Deepak, K. K., Manchananda, S. K., & Maheswari, M. C. (1994). Meditation improve clinicoelectroencephalographic measures in drug-resistant epileptics. *Biofeedback & Self-Regulation, 19*(1): 25–40.

Detert, N., & Douglas, L. (2014). Mindfulness MBSR/MBCT improves general psychiatric symptom severity, depression, anxiety and perceived stress in neurological disorders in an NHS clinical setting. *Neuro-Disability & Psychotherapy, 2*(1/2): 137–156.

Gemmell, C., & Leathem, J. M. (2006). A study investigating the effects of tai chi chuan: individuals with traumatic brain injury compared to controls. *Brain Injury, 20*: 151–156.

Grossman, P., Kappos, L., Gensicke, H., D'Souza, M., Mohr, D., Penner, I., & Steiner, C. (2010). MS quality of life, depression and fatigue improve after mindfulness training. *Neurology, 75*: 1141–1149.

Hackney, M. E., & Earhart, G. M. (2008). Tai Chi improves balance and mobility in people with Parkinson disease. *Gait & Posture, 28*: 456–460.

Hart, J., Kanner, H., Gilboa-Mayo, R., Haroeh-Peer, O., Rozenthul-Sorokin, N., & Eldar, R. (2004). Tai Chi Chuan practice in community-dwelling persons after stroke. *International Journal of Rehabilitation Research, 27*: 303–304.

Husted, C., Pham, L., Hekking, A., & Niederman, R. (1999). Improving quality of life for people with chronic conditions: the example of t'ai chi and multiple sclerosis. *Alternative Therapies in Health & Medicine, 5*: 70–74.

Jahnke, R., Larkey, L., Rogers, C., Etnier, J., & Lin, F. (2010). A comprehensive review of health benefits of qi gong and tai chi. *American Journal of Health Promotion, 24*(6): e1–e25.

Johansson, B., Bjhur, H., & Rönnbäck, L. (2012). Mindfulness-based stress reduction (MBSR) improves long-term mental fatigue after stroke and traumatic brain injury. *Brain Injury, 26*: 1621–1628.

Kabat-Zinn, J. (2003). Mindfulness-based interventions in context: past, present and future. *Clinical Psychology: Science & Practice, 10*: 144–156.

Kaltenmark, M. (1969). *Lao Tzu and Taoism* (R. Greaves trans.). Stanford: Stanford University Press.

Kim, D. K., Lee, K. M., Kim, J., Whang, M. C., & Kang, S. W. (2013). Dynamic correlations between heart and brain rhythm during autogenic meditation. *Frontiers in Human Neuroscience, 7*: 414.

Lee, B. (2000). *Striking Thoughts: Bruce Lee's Wisdom for Daily Living*. New York: Tuttle Publishing.

Li, F. (2013). Tai Ji Quan exercise for people with Parkinson's disease and other neurodegenerative disorders. *International Journal of Integrative Medicine, 1*(4): 1–7.

Li, F., Harmer, P., Fisher, K. J., Junheng, Xu., Fitzgerald, K., & Vongjaturapat, N. (2007). Tai Chi-based exercise for older adults with Parkinson's disease: a pilot program evaluation. *Journal of Aging & Physical Activity, 15*: 139–151.

Li, F., Harmer, P., Fitzgerald, K. J., Eckstrom, E. Stock, R., Galver, J., Maddalozzo, G., & Batya, S. S. (2012). Tai chi and postural stability in patients with Parkinson's disease. *New England Journal of Medicine, 366*: 511–519.

Little, J., & Lee, B. (2000). *Bruce Lee: A Warrior's Journey*. (TV Documentary). Warner Home Video.

Lundgren, T., Dahl, J., Yardi, N., & Melin, L. (2008). Acceptance and commitment therapy and yoga for drug-refractory epilepsy: a randomized controlled trial. *Epilepsy & Behaviour, 13*(1): 102–108.

Luria, A. R. (1975). *The Man with a Shattered World*. Harmondsworth: Penguin.

Lynton, H., Kligler, B., & Shiflett, S. (2007). Yoga in stroke rehabilitation: a systematic review and results of a pilot study. *Topics in Stroke Rehabilitation, 14*(4): 1–8.

Lyubimov, N. N. (1998). Changes in electroencephalogram and evoked potentials during application of the specific form of physiological training (meditation). *Human Physiology, 25*: 171–180.

McCraty, R., Atkinson, M., Tiller, W. A., Rein, G., & Watkins, A. (1995). The effects of emotions on short term heart rate variability using power spectrum analysis. *American Journal of Cardiology, 76*: 1089–1093.

Merton, T. (1969). *The Way of Chuang Tzu*. New York: New Directions.

Mills, N., Allen, J., & Carey-Morgan, S. (2000). Does Tai Chi/Qi Gong help patients with Multiple Sclerosis? *Journal of Bodywork and Movement Therapies, 4*(1): 39–48.

Quinn, D. (unpublished). *The Tai Chi Movement for Wellbeing Effectiveness Measure*.

Quinn, D., & Jones, K. (2012). Tai chi movement (TMW) and embodied mindfulness in mental and physical health. *Inservice Presentation for Herefordshire Primary Care Trust, UK*.

Rajesh, B., Jayachandran, D., Mohandas, G., & Radhakrishnan, K. (2006). A pilot study of a yoga meditation protocol for patients with medically refractory epilepsy. *The Journal of Alternative and Complementary Medicine*, 12(4): 367–371.

Russell, T. A. (2011). Body in mind training: mindful movement for severe and enduring mental illness. *British Journal of Wellbeing*, 2(4): 13–16.

Russell, T. A., & Tatton-Ramos, T. (2014). Body in mind training: mindful movement for the clinical setting. *Neuro-Disability & Psychotherapy*, 2(1/2): 108–136.

Sacks, O. (1973). *Awakenings*. New York: Random House.

Schaffer, O. (2013). *Crafting Fun User Experiences: A Method to Facilitate Flow*. Fairfield, IA: Human Factors International.

Shapira, M. Y., Chelouche, M., Yanai, R., Kaner, C., & Szold, A. (2001). Tai Chi Chuan practice as a tool for rehabilitation of severe head trauma: 3 case reports. *Archives of Physical Medicine & Rehabilitation*, 82(9): 1283–1285.

Shravat, A. (2014). The role of a Yoga group within a holistic rehabilitation setting for ABI using compassion focused therapy approach. A qualitative case illustration. *Neuro-Disability & Psychotherapy*, 2(1/2): 100–107.

Tavee, J., Rensel, M., Planchon, S. M., Butler, R. S., & Stone, L. (2011). Effects of meditation on pain and quality of life in multiple sclerosis and peripheral neuropathy. *International Journal of MS Care*, 13: 163–168.

Taylor-Piliae, R. E., & Coull, B. M. (2012). Community-based Yang-style Tai Chi is safe and feasible in chronic stroke: a pilot study. *Clinical Rehabilitation*, 26(2): 121–131.

Taylor-Piliae, R. E., & Haskell, W. L. (2007). Tai Chi exercise and stroke rehabilitation. *Topics in Stroke Rehabilitation*, 14(4): 9–22.

Venglar, M. (2005). Tai chi and parkinsonism. *Physiotherapy Research International*, 10(2): 116–121.

Wang, C., Bannuru, R., Ramel, J., Kupelnick, B., Scott, T., & Schmid, C. H. (2010). Tai Chi on psychological well-being: systematic review and meta-analysis. *BMC Complementary Alternative Medicine*, 10: 23.

Wang, C., Collet, J. P., & Lau, J. (2004). The effect of tai chi on health outcomes in patients with chronic conditions: a systematic review. *Archives of Internal Medicine*, 164: 493–501.

Wei, G. X., Dong, H. M., Yang, Z., Luo, J., & Zou, X. N. (2014). TaiChiChuan optimizes the functional organization of the intrinsic human brain architecture in older adults. *Frontiers in Aging Neuroscience, 6*: 1–10.

Wei, G. X., Xu, T., Fan, F. M., Dong, H. M., Jiang, L. L., Li, H. J., Yang, Z., Luo, J., & Zuo, X. N. (2013). Can taichi reshape the brain? A brain morphometry study. *PLos One, 8*(4): e61038.

Wilson, B. A., Evans, J., Baddeley, A., & Shiel, A. (2010). Errorless learning in the rehabilitation of memory impaired people. *Neuropsychological Rehabilitation, 4*(3): 307–326.

Yeates, G. N., & Farrell, G. (2014). Editorial for special issue: "Accepting, soothing and stilling cluttered and critical minds in neurological conditions: the influence of Eastern practices". *Neuro-Disability & Psychotherapy, 2*(1/2): 1–2.

Yeates, G. N., Murphy, M., Baldwin, J., Wilkes, J., & Mahadevan, M. (2015). A pilot evaluation of a yoga group for survivors of acquired brain injury in a community setting. *Clinical Psychology Forum, 267*: 17–20.

Zhang, L., Layne, C., Lowder, T., & Liu, J. (2011). A review focused on the psychological effectiveness of Tai Chi on different populations. *Evidence-based Complementary & Alternative Medicine, 2012*: 1–9.

CHAPTER NINE

Mindfulness MBSR/MBCT in a UK public health neurological service: depression, anxiety, and perceived stress outcomes in a heterogeneous clinical sample of ninety-eight patients with neurological or functional neurological disorders*

Niels Detert and Laura Douglass

H ere we report the outcomes of using group mindfulness training as an intervention to address stress, anxiety, depression, and coping in patients with diverse conditions such as multiple sclerosis (MS), epilepsy, brain tumours, or functional disorders, referred to psychology within a UK public sector, National Health Service (NHS) regional neurology and neurosurgery centre. This mindfulness service grew out of the recognition of the unmet psychological needs of neurosciences patients in the context of limited resources. In planning a response to this service need, mindfulness training apparently offered solutions to the key problems: first, it is a group intervention and can be delivered to larger numbers with less clinician time than one-to-one interventions; second, it has a good evidence base; third, it has proved useful with diverse health and mental health patient groups, allowing a single mindfulness programme to be offered to a patient group with diverse symptoms, including neurological dysfunction and disability, depression, anxiety, pain, functional symptoms, and in some cases co-morbid complex psychiatric syndromes.

* Originally published in 2014 in *Neuro-Disability & Psychotherapy*, 2(1/2): 137–156.

In broad terms the evidence for the effectiveness of mindfulness training is already established, both with randomised controlled trials in particular clinical groups (e.g., Ma & Teasdale, 2004; Teasdale et al., 2000), and with meta-analyses of studies in diverse clinical populations (Grossman et al., 2004; Hoffman et al., 2010).

Mindfulness in this population has been understudied, and in the NHS provision of mindfulness-based cognitive therapy (MBCT) has developed more rapidly in mental health settings than in physical health settings, so that mindfulness training is a relatively unfamiliar intervention to neurosciences physicians and professionals. There have nevertheless been a limited number of studies in neurological populations. Of particular relevance to the current study is a randomised controlled trial (RCT) in 150 patients with MS, which demonstrated improved quality of life, depression, and fatigue (Grossman et al., 2010). Other than this our literature search found only two small studies with patients with acquired brain injury: a randomised controlled trial with fifteen patients in a mixed stroke and traumatic brain injury sample that demonstrated improved fatigue (Johansson et al., 2012), and an uncontrolled study in twenty patients with traumatic brain injury that found improved depression, pain, and energy (Bedard et al., 2012).

Neurological patients share some similar stressful experiences with other health groups, and in the absence of more evidence with neurological populations, studies of mindfulness in patients with cancer and patients with pain are also of interest. There is extensive evidence for reduction of psychological symptoms in cancer populations, in a large RCT (Wurtzen et al., 2012), and in meta-analyses (Cramer et al., 2012; Piet et al., 2012; Zainal et al., 2012). In pain patients two reviews have concluded that there is evidence for reduced psychological symptoms and pain intensity (Chiesa & Serretti, 2011; Reiner et al., 2012), although there is still debate about the possibility that this is due to non-specific effects.

Of potential interest for patients with cognitive impairment is evidence that shows cognitive improvements from mindfulness training (e.g., Chiesa et al., 2011; Geng et al., 2011; Jensen et al., 2012; Jha et al., 2007; van den Hurk et al., 2010; Zeidan et al., 2010). However, this research has all been in non-neurological subjects, and so the possibility for improved cognition in people with cognitive impairment remains hypothetical, although an obvious subject for future research.

Taken together this body of evidence provides adequate justification for using mindfulness in a neurological population to help manage depression, anxiety, and stress, to improve coping, and improve pain management. However, in the case of patients with functional neurological symptoms there is no published evidence for the use of mindfulness training. There is some indirect evidence that can support it. For example, there is evidence in patients with somatisation disorders such as fibromyalgia, chronic fatigue syndrome, and irritable bowel syndrome of small to moderate effect sizes for improving pain, depression, anxiety, and quality of life in a systematic review and meta-analysis of RCTs of mindfulness training (Lakhan & Schofield, 2013). There is also other indirect evidence in the form of pilot studies of one-to-one and group psychological approaches showing reduced seizure frequency in patients with non-epileptic attacks (Baslet, 2012). Further indirect evidence comes from the use of mindfulness as a component of dialectical behaviour therapy with personality disorder (e.g., Lynch & Bronner, 2006; Wagner et al., 2006) with the aims of improving emotional regulation and reducing reactive, impulsive, or inflexible responding. In general people with functional neurological symptoms have problems with emotion awareness and regulation, and improved emotion regulation is a key target and outcome of mindfulness training (e.g., Britton et al., 2012; Farb et al., 2012a,b; Robins et al., 2012; Taylor et al., 2011). Based on these considerations there is a twofold theoretical justification for using mindfulness with people with functional symptoms, especially non-epileptic attacks: first, functional symptoms are stress triggers in themselves like other symptoms and improving coping with this is a treatment aim in itself; second, improved emotion and anxiety regulation may reduce functional symptoms.

Mindfulness training can be thought of as a particular form of attention training. The basic procedure, repeated several times during periods of practise, is to attend to aspects of immediate experience such as sensory experiences. This mode of simple awareness of here-and-now experience is contrasted with a non-mindful state, "automatic pilot", and a "driven-doing mode" (e.g., Segal et al., 2013, p. 68). In "automatic pilot" attention is easily captured, ruminative thinking may govern feelings, and automatic reactions govern behaviour. In the "driven-doing mode" inner experiences become the objects of attempts at control with resulting self-critical or self-defeating patterns. More comprehensively, mindfulness has been defined as "the awareness that emerges through paying attention on purpose, in the present moment, and non-judgementally to the unfolding of experience moment to moment" (Kabat-Zinn, 2003, p. 145). The importance of a self-compassionate quality to this non-judgemental awareness has been emphasised by recent studies (e.g., Kuyken et al., 2010).

Mindfulness is thought of as a natural capacity, which is underdeveloped and utilised, but that can be trained. Training the capacity to be aware of experiences (inner and outer) in real time, and relating to these experiences non-judgementally, and self-compassionately, enables a calmer, less habit-driven and more conscious and adaptive response to the experiences of life and in particular the difficult and aversive experiences, such as physical pain and illness, anxiety, depressive thoughts, intense emotion, impulses to self-defeating behaviours or self-destructive impulses, and difficult mind-states. The skill of mindfulness is trained in two main ways: first, by practising meditation, in other words practising awareness of experience and observing the mind's activity during periods deliberately set aside, and second, by practising the same mindful awareness during everyday life. Implicit in the standard practise are reduction of functional avoidance behaviour and exposure to difficult experience, including inner avoidance such as distraction or day-dreaming, and reduction of reactive rumination, by awareness of experience and awareness of thinking, fostering anxiety regulation, and adaptive responding.

Mindfulness training is usually offered in an eight-week group format modelled on MBCT (Segal et al., 2013), or mindfulness-based stress reduction (MBSR; e.g., Kabat-Zinn, 1982). The programme usually consists of eight weekly sessions of about two hours (range, 1.5–2.5 hours), including periods of teacher-guided mindfulness practise, group discussion, various in-class and homework short exercises, with a significant home practice component of forty to sixty minutes daily (using audio-guided practises), and mindfulness exercises in everyday life, as well as a little background reading. There is a whole-day five to seven hour session in week six, omitted in some programmes.

The main goal of this study was to evaluate the clinical usefulness of the mindfulness programme in a neurological population. Based on previous research the main predicted outcomes were in psychiatric symptoms. A further predicted outcome was of increased capacity to cope with difficulties and stress, and this is of particular interest in a population with long-term conditions that present continuing and in some cases increasing stressors in the form of neurological symptoms, impairment, and disability. Although the present study in itself does not have the randomised controlled design needed to draw strong conclusions about treatment effectiveness, the effectiveness of mindfulness is well-established by previous research, and the question becomes whether, in a neurological population, it shows clinical effects on psychological measures comparable with reported effect sizes in the literature. Based on a meta-analysis (Hoffman et al., 2010) of studies in heterogeneous populations, uncontrolled pre-post effect sizes of medium size would be predicted (anxiety, Hedges g = 0.63; depression, Hedges g = 0.59).

The study

Participants

The study reports on the outcome data of a convenience sample, collected as part of a clinical mindfulness service over a four year period. The ninety-eight participants whose results are reported here were referred to the mindfulness service by neurologists, clinical nurse specialists, and general practitioners, with stress, depression, anxiety, or other psychiatric symptoms. Questionnaire measures were given at initial interview, up to eight weeks pre-course, and again at the end of the eight week course. All participants with data from both assessment points were entered into this analysis.

There was significant attrition of participant numbers due to some drop-out from the course and some non-completion of post-course questionnaires. 182 people enrolled on the course in this period and 147 completed it (81%). Reasons for non-completion were identifiable for twenty-three out of the thirty-five who dropped out: seven had practical difficulties such as transport, seven suffered illness, six thought the course was not useful to them, two experienced symptom improvement and discontinued, one did not attend the first session, and for twelve there was no reason given. This represents a 19% drop-out rate. These figures and drop-out rate are consistent with those of Wurtzen and colleagues (2012) who found that one third dropped out due to disappointment with mindfulness, one third due to transport/

scheduling problems, and one third due to sickness. Of the 147 people who completed the course, forty-eight did not hand in questionnaires, and one withheld consent to use their data. It has not been possible retrospectively to determine the reasons for questionnaire non-completion except in eight cases who missed their last session for logistical reasons such as snow or transport difficulties. This means that for 27% of the original sample there was no recorded reason for the lack of post-course data. It was the first author's impression that this was due to ordinary forgetfulness as participants were asked to complete their questionnaires at home, but of course it is impossible to exclude any systematic effects on this. Finally, questionnaires were available for analysis from ninety-eight (67%) participants.

The diagnoses and demographic data are summarised in Table 1. The service is for adult patients (age eighteen or more) of the neuro-sciences service, including patients with neurological disorders/conditions, and patients with functional neurological symptoms (FNS). Inclusion criteria were broad, including severe physical disability, moderate cognitive impairment, and severe mental health problems. Such problems as suicidality, active post-traumatic stress disorder (PTSD), and complex needs/personality disorder were not usually a cause for exclusion, although where appropriate concurrent referrals to other services were arranged. Exclusion was rare and based on clinical judgement that the course would be impossible for a patient to make use of, or cause disadvantage or risk to other group members (e.g., severe social phobia, active psychosis, risk to others, or high risk to self). In practice, a smaller number had severe mobility problems, moderate cognitive impairment, or severe mental health symptoms. A significant minority had FNS in the form of non-epileptic attacks or functional motor symptoms. A greater number of participants were ambulatory (some using a walking aid), with moderate to severe depression and anxiety, mild or minimal cognitive impairment, and with a variety of other neurological symptoms, pain, and fatigue.

There is a notably larger proportion of female participants in the sample, with a female to male ratio of 2.5:1. This seems broadly consistent with the known 2.5:1 female to male ratio in the epidemiology of depression (NIHCE, 2003), and in MS in which there is a 2.3:1 ratio (Alonso & Hernan, 2008).

This population reflects the clinical setting, which is a regional outpatient neurosciences centre, incorporating neurology, neurosurgery,

Table 1: Demographic data.

Demographic	Number
Diagnosis:	
Multiple sclerosis	36
Functional neurological symptoms	16
Brain tumour	13
Epilepsy	10
Cerebrovascular	5
Traumatic brain injury	3
Migraine	2
Parkinson's disease	2
Transverse myelitis	2
Other neurological	3
Other medical	3
Other psychiatric	3
Total	98
No. of males	28 (29%)
No. of females	70 (71%)
Age: mean (SD	44.0 (12.0)
Age range	18–73

and neuropsychology, and not a neuro-rehabilitation service, in which there would be a larger proportion of traumatic brain injury and other acquired brain injury patients.

The sample is very heterogeneous, and this may mask sub-group differences. Despite the heterogeneity there are some identifiable larger sub-groups of fifteen or more members. In particular the sub-group of functional patients have different characteristics in that they may not have neurological conditions, and have a greater likelihood of more severe and persistent psychological dysfunction. Within the neurological conditions there are also progressive conditions and a variety of other non-progressive conditions, and these are also clinically meaningful sub-groups. Although the sample as a whole reflects routine clinical practice in our service, the overall group outcome may not be representative of these sub-groups' outcomes. We therefore further stratified the whole sample for a secondary analysis into these three sub-groups: Functional (n = 16), neurological non-progressive (n = 30), neurological progressive (*n* = 46). The functional sub-group included sixteen participants with functional neurological symptoms, and these can be further broken down: dissociative seizures or non-epileptic attacks (9), functional weakness or movement disorder (5), functional memory problems (2). The neurological non-progressive group remained quite heterogeneous and included people with traumatic brain injury (3), cerebrovascular injury (5), epilepsy (10), non-progressive brain tumours (6), transverse myelitis (3), migraines (2), and Tourette's syndrome (1). The neurological progressive sub-group included people with MS (36), progressive brain tumours (8), and Parkinson's disease (2). Six patients, with purely psychiatric or non-neurological medical conditions could not be categorised in these three groups and were omitted from these analyses.

Intervention

Mindfulness training was provided in a standard eight-week group format, with groups varying in size between ten to twenty participants, with the same instructor (the first author). Sessions were two and a half hours long. There was an additional whole-day session on a Saturday between sessions six and seven of the course. Three courses were run per year, and the results from thirteen consecutive cycles of mindfulness courses are presented here.

The curriculum underwent revision between cycle eleven and twelve. The first eleven cycles were run using a modified MBSR curriculum with elements of MBCT, and cycles twelve to thirteen shifted to a modified MBCT curriculum, using slightly edited versions of the standard MBCT handouts (Segal et al., 2013), which address long-term conditions rather than recurrent depression. As significant elements of MBCT were already included in the modified MBSR curriculum, the activities and practises on the courses were similar, and the core mindfulness practices were the same. The main changes were the handouts, and some slight changes to the order of topics in the course. All courses used the same mindfulness practices during sessions and the same audio-guided recordings of the core mindfulness practises.

Measures

Brief symptom inventory (Derogatis, 1993): the BSI is a fifty-three-item short form of the SCL-90-R. It is a broad spectrum measure of psychiatric symptoms with nine subscales (somatic symptoms, obsessive–compulsive, interpersonal sensitivity, depression, anxiety, hostility, phobic anxiety, paranoid ideation, psychoticism), as well as a total score (general severity index, GSI). It uses a five-point Likert scale (0–4), and subscale and GSI scores are expressed as means of the single item scores. It has good psychometric properties (Derogatis & Melisaratos, 1983). Items include, for example, "feeling no interest in things", "feeling so restless you couldn't sit still", and "feeling that you are watched or talked about by others".

Perceived stress scale (Cohen et al., 1983): the PSS is a ten-item questionnaire designed to measure the degree to which life situations are perceived as stressful. It has adequate psychometric properties (Cohen et al., 1983). It uses a five-point Likert scale (0–4). Examples of items include "In the last month, how often have you been upset because of something that happened unexpectedly?", and "In the last month, how often have you felt confident about your ability to handle your personal problems?"

Results

Data analysis

The analysis used SPSS version 20. Tests for normality, skewness, and kurtosis were satisfactory for the main measures (BSI, GSI, and PSS). Two out of nine subscales of the BSI showed excess skewness (hostility and phobic anxiety), and one subscale showed excess kurtosis (hostility). For statistical analyses a logarithmic data transformation (Log 10; constant, +1.0) was used on the hostility and phobic anxiety subscales of the BSI. Untransformed means, standard deviations, and effect sizes for the sample as a whole are reported in Table 2.

The main analysis tested the prediction of within-subject change from pre-course assessment to post-course assessment in the whole sample. A one factor within-subjects ANOVA was used. This showed significant change with large effect sizes (partial eta squared, ηp^2) on the BSI GSI ($F(1, 97) = 110.741$, $p < 0.001$, $\eta p^2 = 0.533$) and PSS ($F(1, 97) = 513.706$, $p < 0.001$, $\eta p^2 = 0.841$). Cohen's d was medium for the BSI GSI (Cohen's $d = 0.74$), and very large for the PSS (Cohen's $d = 2.28$).

One factor within-subjects ANOVAs were also performed on the BSI subscales. All nine subscales showed significant change with large

Table 2: Whole sample means, standard deviations and effect sizes for each measure before and after mindfulness training (MT)

Measure	Pre MT mean (SD)	Post MT Mean (SD)	Partial eta squared	Cohen's d
Brief symptom inventory				
General severity index	1.33 (0.68)	0.84 (0.64)***	0.533	0.74
Perceived stress scale	2.35 (0.56)	1.14 (0.50)***	0.841	2.28
BSI subscales				
Somatisation	1.24 (0.82)	0.90 (0.70)***	0.322	0.45
Obsessive –compulsive	2.15 (1.04)	1.45 (0.91)***	0.468	0.72
Interpersonal sensitivity	1.52 (1.00)	0.87 (0.90)***	0.389	0.68
Depression	1.41 (0.90)	0.81 (0.79)***	0.387	0.71
Anxiety	1.48 (0.95)	0.86 (0.81)***	0.427	0.70
Hostility	1.09 (0.86)	0.63 (0.68)***	0.371	0.59
Phobic anxiety	1.01 (1.02)	0.67 (0.93)***	0.242	0.35
Paranoid ideation	1.02 (0.87)	0.69 (0.75)***	0.185	0.41
Psychoticism	0.97 (0.80)	0.54 (0.64)***	0.328	0.59

Note: ***$p < 0.001$

effect sizes (partial eta squared): somatisation ($F(1, 97) = 45.997$, $p < 0.001$, $\eta p^2 = 0.322$), obsessive–compulsive ($F(1,97) = 85.251$, $p < 0.001$, $\eta p^2 = 0.468$), interpersonal sensitivity ($F(1, 97) = 61.658$, $p < 0.001$, $\eta p^2 = 0.389$), depression ($F(1, 97) = 61.303$, $p < 0.001$, $\eta p^2 = 0.387$), anxiety ($F(1, 97) = 72.294$, $p < 0.001$, $\eta p^2 = 0.427$), hostility ($F(1, 97) = 67.02$, $p < 0.001$, $\eta p^2 = 0.409$), phobic anxiety ($F(1, 97) = 36.90$, $p < 0.001$, $\eta p^2 = 0.276$), paranoid ideation ($F(1, 97) = 22.070$, $p < 0.001$, $\eta p^2 = 0.185$), psychoticism ($F(1, 97) = 47.347$, $p < 0.001$, $\eta p^2 = 0.328$). Cohen's *d* effect sizes are shown in Table 2, and are medium ($d > 0.5$) for depression, anxiety, interpersonal sensitivity, obsessive–compulsive, hostility, and psychoticism, and small ($d > 0.2$) for somatisation, phobic anxiety, and psychoticism.

The question of whether there were significant differences between the three sub-groups (functional, neurological-progressive, and neurological non-progressive) was addressed with a two factor mixed

ANOVA with time and group as factors. There was a significant interaction of time and group on the GSI (F(2,89) = 6.790, p = 0.002, ηp^2 = 0.132), but not on the PSS (F(2,89) = 2.631, p = 0.078, ηp^2 = 0.056), although there is a trend.

A further two factor mixed ANOVA was performed to test the hypothesis that the two neurological sub-groups (progressive and non-progressive) did not differ, with time and group as factors. This found that there was in fact a significant interaction of time and group on the GSI with medium effect size (F(1, 74) = 4.747, p = 0.033, ηp^2 = 0.060), but not on the PSS (F(1, 74) = 1.926, p = 0.169, ηp^2 = 0.025).

Subsequent one factor within-subjects ANOVAs were carried out for each of the three sub-groups on the main measures. These reveal significant within-subject changes from pre-course to post-course on both GSI and PSS in all three groups, but a smaller effect size and significance only at the 5% level in the functional group on the GSI: functional GSI (F(1,15) = 6.022, p = 0.027, ηp^2 = 0.286), functional PSS (F(1, 18) = 28.916, p < 0.001, ηp^2 = 0.658), neurological non-progressive GSI (F(1, 29) = 21.431, p < 0.001, ηp^2 = 0.425), neurological non-progressive PSS (F(1, 29) = 160.542, p < 0.001, ηp^2 = 0.847), neurological progressive GSI (F(1, 45) = 96.126, p < 0.001, ηp^2 = 0.681), neurological progressive PSS (F(1, 45) = 359.753, p < 0.001, ηp^2 = 0.889). Calculating Cohen's d for each subgroup reveals that on the GSI the functional subgroup show only a small effect size (d = 0.20), the neurological non-progressive group show a medium effect size (d = 0.62), and the neurological progressive group show a large effect size (d = 1.13). All sub-groups showed very large effect sizes on the PSS (functional, d = 1.41; neurological non-progressive, d = 1.91; neurological progressive, d = 2.96). The sub-group means, standard deviations, and effect sizes are summarised in Table 3.

Discussion

The present findings support the effectiveness of mindfulness training as an intervention to reduce general psychiatric symptoms, depression, and anxiety, and increase stress coping capacity in a neurological population. In the sample as a whole we found medium pre-post effect sizes on symptoms of depression and anxiety, consistent with those in the literature. Stress coping (perceived stress scale) was significantly

Table 3: Diagnostic sub-group means, standard deviations and effect sizes for each measure before and after mindfulness training (MT)

Group	N	Measure	Pre MT mean (SD)	Post MT mean (SD)	Partial eta squared	Cohen's d
Functional	16	GSI	1.46 (0.81)	1.26 (0.92)*	0.286	0.2
	16	PSS	2.28 (0.75)	1.30 (0.64)***	0.658	1.41
Neurological	30	GSI	1.18 (0.71)	0.77 (0.62)***	0.425	0.62
non-progressive	30	PSS	2.31 (0.66)	1.15 (0.55)***	0.847	1.91
Neurological	46	GSI	1.37 (0.63)	0.73 (0.49)***	0.681	1.13
progressive	46	PSS	2.38 (0.46)	1.06 (0.43)***	0.889	2.96

Note: GSI: brief symptom inventory general severity index, PSS: perceived stress scale.
*$p < 0.05$, ***$p < 0.001$.

improved with a very large effect size, and is one of the more interesting findings and very relevant for a neurological population with a need to cope with symptoms, impairment, and disability in the long-term.

The uncontrolled, non-randomised pre-post design of the study is a limitation in principle on the interpretation of the results. If this were an isolated study of mindfulness it would be hard to draw conclusions about effectiveness of the intervention with confidence. However, in the context of previous evidence of the effectiveness of mindfulness as an intervention from RCTs and meta-analyses, the goals of this study were to evaluate whether similar effects as seen in other populations could be demonstrated in a clinical sample in a neurological setting. These findings do show similar effect sizes to those previously found with stronger methodologies in the literature. Accordingly, while this design does not allow a strong conclusion about the effectiveness of mindfulness training from these results alone, the results do support the weaker conclusion that the observed uncontrolled pre-post effect size in a neurological population is consistent with meta-analytic results of pre-post designs in other populations (Hoffman et al., 2010), and this does provide a reasonable degree of confidence that the intervention is effective in the population and setting studied.

A number of other limitations affect these findings. The main weaknesses of the study are in the drop-out rate and questionnaire completion rate. Better information is available on reasons for drop-out from the course. For those participants for whom information was available about a third dropped out for logistical reasons, a third because of illness, and a third were disappointed with the course. There were also some drop-outs for whom no reason was known, but it may be reasonable to suppose that their reasons may have fallen in the same three categories in the same proportions, which is supported by the observation that the proportions in our study were the same as previously reported in another study with a similar drop-out rate of about 20% (Wurtzen et al., 2012). For those who were disappointed with mindfulness, presumably there is a higher likelihood that their symptoms would not have improved, and so the fact that they dropped out would tend to increase the final effect size of the course. In drop-outs for logistical reasons, or because of illness, it may be more reasonable to assume no systematic effects and so their outcome profile might not be significantly different from the rest of the sample. Accordingly the net effect of the drop-outs may have been to slightly exaggerate the effect sizes above what they would have been if all of the enrolled participants had completed the course, although it seems unlikely that the 6–7% who found the course unhelpful would have very significantly changed the overall results.

It is harder to interpret the effects of the missing data from course completers. 33% of 147 course completers did not hand in post-course questionnaires. This raises the possibility that those who did not complete final questionnaires may have had a different outcome profile, as mood might be a factor in remembering to complete and hand in forms. For a few of the non-completers there are known logis-tical reasons, but for 27% of the original sample of course completers the reasons are unknown, and, although there was no reason to suspect anything more than ordinary forgetfulness, it is not possible to exclude systematic effects that might have negatively affected mean questionnaire scores. On balance it seems more likely that the lack of this data would tend, if anything, to exaggerate effect size than to reduce it. The fact that the present study's findings were of the same order as reported in the literature, and indeed slightly above the meta-analytic effect sizes, tends to support our study's findings and also leaves room for some slight downward adjustment. Based on this one can speculate that it seems unlikely that the missing data would skew the final effect sizes very significantly, but the size of a possible effect is, of course, unknown, and this reduces the certainty in the present study's results.

The heterogeneity of the sample was both a weakness and a strength of this study. On the one hand a heterogeneous population reflects clinical reality, and is unlikely to exaggerate treatment effects, but rather the opposite. On the other hand it is harder to generalise these results to other populations. The secondary analyses on sub-groups of the overall sample are useful to address this and showed significant differences between all three sub-groups on psychiatric symptom outcomes. The functional symptoms sub-group only experienced a small effect on psychiatric symptoms, whereas a neurological group with non-progressive conditions experienced a medium effect size, and a group with progressive neurological conditions experienced a large effect size. Interestingly stress coping showed very large effect sizes in all sub-groups, including the functional symptoms group, and there was no difference between the groups on the perceived stress scale.

The sub-group analyses deserve some further discussion as they show some substantially different results in the three sub-groups. The easiest finding to interpret is that patients with functional symptoms do not do as well as the others on psychiatric symptoms. On the other hand there is evidence that they may experience a small effect on psychiatric symptoms, and do experience a significant and large effect in reduced perceived stress. Of course the numbers in the functional sub-group were small and this finding must be considered preliminary. It nevertheless seems a useful preliminary result, which indicates both that there is probably benefit for this traditionally difficult to help group, and that the symptomatic benefit is not as large as for other patient groups, so that it is reasonable to be more modest about expectations of outcome in functional patients, and to think of mindfulness training either as an adjunct to other treatments or within a purely coping model for this patient group. There were also examples of reduced frequency and severity of non-epileptic seizures in individual cases, but this information was not collected systematically, and so in future it will be important to collect data on the severity of functional symptoms themselves before and after mindfulness training. Further research with a larger group would be useful.

The difference in outcomes between the two neurological groups is harder to interpret. One might make the case for amalgamating these

groups under the umbrella of neurological disorders, but this is not justified in view of the evidence of a difference. The non-progressive
group shows a medium effect size consistent with expectations from
the literature, but the progressive neurological group show a large
effect, nevertheless consistent with the upper bound of the range of previously reported effect sizes. Examining the means of the two groups
before and after mindfulness training reveals that the difference may
mainly reflect a higher level of symptoms in the progressive group at
the first assessment, and larger standard deviations in the smaller non-
progressive group. One possible interpretation therefore is that the progressive neurological conditions sub-group experienced a larger effect
because they had worse symptoms at the outset, and this would be consistent with the findings of Grossman and colleagues (2010) in MS
patients. Another possible interpretation is that the membership of the
progressive group may be a factor as patients with MS dominated in
that sub-group, and there are reasons to hypothesise that mindfulness
might be particularly helpful in MS, that is, people with MS suffer multiple symptoms including sensory symptoms, motor symptoms, pain,
bladder and bowel dysfunction, cognitive dysfunction, and fatigue, in
addition to depression and anxiety, all of which may be mitigated by
mindfulness training based on previous evidence.

The drop-out rate of about 20% is of relevance, as this is a possible
outcome of the course for some participants. The course presents
some practical difficulties to participants, which may be exacerbated
for neurological patients. Neurological patients frequently have
mobility and travel problems and getting to the course can be difficult.
Some patients are more vulnerable to illness or to episodes of neurological symptoms interfering with attendance. These factors lead to
missed sessions for some people, and when these are consecutive it
can be hard to maintain the momentum of the course.

The results also confirm that the eight-week MBSR/MBCT format is
applicable with and acceptable to neurological patients. Relatively
minimal adjustments needed to be made to the course structure,
contents, and delivery, although this patient group does present some
particular issues. The following comments about adjustments in working with cognitive impairment, physical disability, fatigue, and functional symptoms may be helpful in implementing similar programmes
elsewhere.

Cognitive function

- The explicit memory demands are low as instructions are simple, most exercises use audio-guidance (live, or on CD), and didactic teaching is de-emphasised in favour of experiential learning. For people with mild to moderate memory impairment the programme therefore does not require significant adaptation, as long as memory is sufficient to support attending sessions and engaging in home practise. If memory for written material is a problem, then a reasonable adaptation is to omit the written material and focus on experiential mindfulness practise and the core themes.

- Mild to moderate impairments of attention/executive function can be worked with using no adaptations or simple adaptations; audio-guided practises work as an effective attention support, and if reminders are needed for practise during the day then normal practical strategies for reminding can be used. Severe impairment, however, could be an obstacle to participation in a standard format course, and may need more tailored teaching and support, with smaller groups (maybe 1:1 in some cases), shorter sessions, and greater between-session support.

- Severe language impairment would often be an obstacle to standard MBSR/MBCT type programmes, and while greater use of forms of mindful movement, such as elements from yoga or t'ai chi/qigong, could be an avenue to explore; it is beyond the scope of this chapter.

Physical function

- Patients with a wide range of physical disabilities can successfully complete the course, including patients using wheelchairs and with limited trunk, neck, and upper limb function.

- Where people are mobile but movement and balance difficulties require use of a stick or frame, adaptations can be made fairly easily; mindful movement can focus on floor-based exercises, some stretches can be done from a seated position, and participants can omit certain movements, substituting other movements or awareness of breathing while lying or sitting. In a group with a large proportion of people with severe physical disability, then mindful movement could be taught with a focus on mindfulness of small and possible movements, or it could be omitted and alternative mindfulness practices focusing on other senses could be substituted.

• In walking meditation people who use a walking aid should also use that, bringing awareness to their movements in the usual way. People who fatigue during such practise can be encouraged to do shorter practises and stop when they need. People who use a self-propelled wheelchair can also participate, bringing awareness to the movements of their arms and upper bodies. For people with little movement and a powered wheelchair, options include substituting sitting with awareness of breathing, or another sensory mindfulness practice, or indeed moving with the wheelchair and attending to physical sensations that are salient, or to the breath, or to vision, or sound.

• Where there is somatosensory impairment, numbness, or no sensation in parts of the body, the question arises of how to manage this during the body scan practise of attending to sensations in different parts of the body. When this is affecting some body parts and not others, the best approach can be to treat this experience in the same way as other body sensations and pain, which is noticing the experience of whatever sensation is present in each body area, including the experience of no-sensation, as this can still be an effective focus for mindfulness. Where somatosensory impairment is very extensive or near-total, options include focusing only on small regions with sensation, or substituting a different focus for mindfulness, such as another sensory modality, for example, sound or vision.

Fatigue

• This can be a particular issue in the all-day session, and can be managed by allowing flexibility in the programme. Practical measures to manage this include limiting the length of the whole day session to five and a half rather than seven hours. Patients can leave at lunch time if necessary. In fact many patients find they can manage a whole day after all. Periods of walking meditation can be kept brief at about five to ten minutes.

• A more subtle adaptation to dealing with fatigue is to give instructions with particular emphasis on non-striving, a core mindfulness principle. In the early stages of the course partici-pants will often play out habitual patterns of exerting excessive effort to control the mind during mindfulness practise. This becomes tiring during a whole day, and potentially exhausting for many patients with neurological conditions, so instructions can emphasise the opportunity to rest, relax, and save energy, and a good opportunity to learn this core mindfulness principle while also managing fatigue better.

Functional neurological symptoms

- The main adaptation for people with FNS is recognition of the propensity for such people to be overwhelmed by emotion-related anxiety and for this to manifest as problematic somatic symptoms. Since the default approach in mindfulness training is to encourage opening to and allowing all kinds of inner and outer experience, this could result in excessive anxiety and exacerbation of symp-toms for such patients in some situations. For example, if there is over-whelming anxiety in relation to re-experiencing of trauma, or an experience of anger in a conflict, this may quickly lead to debil-itating symptoms, and encouraging opening to the experience of anger or traumatic memory might lead to intolerable increases in anxiety when anxiety is already severely limiting functioning.
- People with this degree of sensitivity need to adopt a more graded approach to exposure to anxiety-provoking experiences; in such situations mindfulness can be used for sensory ground-ing and anxiety regulation. A very effective focus is the bodily experience of anxiety itself, or alternatively a neutral sensory experience, and not opening to the anxiety-provoking experience (e.g., traumatic memory, angry feeling), until anxiety comes down again.
- Reprocessing of traumatic memories is best done through psychological therapy, not as part of a mindfulness course.
- Exposure to anxiety-provoking emotions can be done through the use of skills learned on a mindfulness course, such as the "working with difficulty" practise, but should be done in a graded way beginning with low intensity.
- In moments of high anxiety or attacks of functional symptoms, people with FNS may need some guidance in focusing on

sensory grounding, especially if there is limited awareness of anxiety, or if there is cognitive disruption due to high anxiety.

- Psycho-education about the bodily symptoms of anxiety and emotion can be very useful in labelling problematic physical experiences accurately, enabling people to begin to develop effective anxiety-regulating responses.

In conclusion, mindfulness training is practical and well-accepted by patients in a mixed neurological and functional group. The findings provide evidence that the previously established benefits of mindfulness training can be extended to a mixed neurological group in a heterogeneous clinical population. Mindfulness training may be a particularly appropriate intervention for people suffering with long-term conditions who have a long-term need to cope with the stress of problematic symptoms, impairments, and disability. For people with functional symptoms the present findings represent preliminary evidence for benefits in stress coping, and suggest further research is justified.

References

Alonso, A., & Hernan, M. (2008). Temporal trends in the incidence of multiple sclerosis: a systematic review. *Neurology*, *71*: 129–135.

Baslet, G. (2012). Psychogenic nonepileptic seizures: a treatment review. What have we learned since the beginning of the millennium? *Neuropsychiatric Disease and Treatment*, *8*: 585–598.

Bedard, M., Felteau, M., Marshall, S., Dubois, S., Gibbons, C., Klein, R., & Weaver, B. (2012). Mindfulness-based cognitive therapy: benefits in reducing depression following a traumatic brain injury. *Advances in Mind-body Medicine*, *26*: 14–20.

Britton, W., Shahar, B., Szepsenwol, O., & Jacobs, W. J. (2012). Mindfulness-based cognitive therapy improves emotional reactivity to social stress: results from a randomized controlled trial. *Behaviour Therapy*, *43*: 365–380.

Chiesa, A., Calati, R., & Serretti, A. (2011). Does mindfulness training improve cognitive abilities? A systematic review of neuropsychological findings. *Clinical Psychology Review*, *31*: 449–464.

Chiesa, A., & Serretti, A. (2011). Mindfulness-based interventions for chronic pain: a systematic review of the evidence. *The Journal of Alternative and Complementary Medicine*, *17*: 83–93.

Cohen, S., Kamarck, T., & Mermelstein, R. (1983). A global measure of perceived stress. *Journal of Health and Social Behaviour, 24*: 385–396.

Cramer, H., Lauche, R., Paul, A., & Dobos, G. (2012). Mindfulness-based stress reduction for breast cancer—a systematic review and meta-analysis. *Current Oncology, 19*: 343–352.

Derogatis, L. (1993). *Brief Symptom Inventory (BSI) Administration, Scoring, and Procedures Manual* (4th edn). Minneapolis, MN: Pearson.

Derogatis, L. R., & Melisaratos, N. (1983). The brief symptom inventory: an introductory report. *Psychological Medicine, 13*: 595–605.

Farb, N., Anderson, A., & Segal, Z. (2012a). The mindful brain and emotion regulation in mood disorders. *Canadian Journal of Psychiatry, 57*: 70–77.

Farb, N., Anderson, A., Mayberg, H., Bean, J., McKeon, D., & Segal, Z. (2012b). Minding one's emotions: mindfulness training alters the neural expression of sadness. *Emotion, 10*: 25–33.

Geng, L., Zhang, L., & Zhang, D. (2011). Improving spatial abilities through mindfulness: effects on the mental rotation task. *Consciousness & Cognition, 20*: 801–806.

Grossman, P., Kappos, L., Gensicke, H., D'Souza, M., Mohr, D., Penner, I., & Steiner, C., (2010). MS quality of life, depression, and fatigue improve after mindfulness training. *Neurology, 75*: 1141–1149.

Grossman, P., Niemann, L., Schmidt, S., & Wallach, H. (2004). Mindfulness-based stress reduction and health benefits. A meta-analysis. *Journal of Psychosomatic Research, 57*: 35–43.

Hoffman, S., Sawyer, A., Witt, A., & Oh, D. (2010). The effect of mindfulness-based therapy on anxiety and depression: a meta-analytic review. *Journal of Consulting and Clinical Psychology, 78*: 169–183.

Jensen, C., Vangkilde, S., Frokjaer, V., & Hasselbach, S. (2012). Mindfulness training affects attention—or is it attentional effort? *Journal of Experimental Psychology: General, 141*: 106–123.

Jha, A., Krompinger, J., & Baime, M. (2007). Mindfulness training modifies subsystems of attention. *Cognitive, Affective, & Behavioural Neuroscience, 7*: 109–119.

Johansson, B., Bjuhr, H., & Rönnbäck, L., (2012). Mindfulness-based stress reduction (MBSR) improves long-term mental fatigue after stroke or traumatic brain injury. *Brain Injury, 26*: 1621–1628.

Kabat-Zinn, J. (1982). An outpatient program in behavioural medicine for chronic pain based on the practice of mindfulness meditation. *General Hospital Psychiatry, 4*: 33–47.

Kabat-Zinn, J. (2003). Mindfulness-based interventions in context: past, present, and future. *Clinical Psychology: Science and Practice, 10*: 144–156.

Kuyken, W., Watkins, E., Holden, E., White, E., Taylor, R., Byford, S., Evans, A., Radford, S., Teasdale, J., & Dalgleish, T. (2010). How does mindfulness-based cognitive therapy work? *Behaviour Research and Therapy, 48*: 1105–1112.

Lakhan, S., & Schofield, K. (2013). Mindfulness-based therapies in the treatment of somatization disorders: a systematic review and meta-analysis. *PLoS ONE, 8*(8): e71834. doi:10.1371/journal.pone.0071834

Lynch, T., & Bronner, L. (2006). Mindfulness and dialectical behaviour therapy (DBT): application with depressed older adults with personality disorders. In: R. Baer (Ed.), *Mindfulness-Based Treatment Approaches: Clinician's Guide to Evidence Base and Applications* (pp. 167–189). Burlington, MA: Academic Press.

Ma, S., & Teasdale, J. (2004). Mindfulness-based cognitive therapy for depression: replication and exploration of differential relapse prevention effects. *Journal of Consulting and Clinical Psychology, 72*: 31–40.

National Institute for Health and Clinical Excellence (NIHCE) (2003). *Depression, NIHCE Guideline, second consultation*. London: National Health Service.

Piet, J., Wurtzen, H., & Zachariae, R. (2012). The effect of mindfulness-based therapy on symptoms of anxiety and depression in adult cancer patients and survivors: a systematic review and meta-analysis. *Journal of Consulting and Clinical Psychology, 80*: 1007–1020.

Reiner, K., Tibi, L., & Lipsitz, J. (2012). Do mindfulness-based interventions reduce pain intensity? A critical review of the literature. *Pain Medicine, 14*: 230–242.

Robins, C., Keng, S-L., Ekblad, A., & Brantley, J. (2012). Effects of mindfulness-based stress reduction on emotional experience and expression: a randomized controlled trial. *Journal of Clinical Psychology, 68*: 117–131.

Segal, Z. V., Williams, J. M. G., & Teasdale, J. D. (2013). *Mindfulness-based Cognitive Therapy for Depression* (2nd edn). New York: Guilford.

Taylor, V., Grant, J., Daneault, V., Scavone, G., Breton, E., Roffe-Vidal, S., Courtemanche, J., Lavarenne, A., & Beauregard, M. (2011). Impact of mind-fulness on the neural responses to emotional pictures in experienced and beginner meditators. *NeuroImage, 57*: 1524–1533.

Teasdale, J., Segal, Z., Williams, M., Ridgeway, V., Soulsby, J., & Lau, M. (2000). Prevention of relapse/recurrence in major depression by mindfulness-based cognitive therapy. *Journal of Consulting and Clinical Psychology, 68*: 615–623.

van den Hurk, P., Giommi, F., Gielen, S., Speckens, E., & Barendregt, H. (2010). Greater attentional processing related to mindfulness meditation. *The Quarterly Journal of Experimental Psychology, 63*: 1168–1180.

Wagner, E. E., Rathus, J. H., & Miller, A. L. (2006). Mindfulness in dialectical behaviour therapy (DBT) for adolescents. In: R. Baer (Ed.), *Mindfulness-Based Treatment Approaches: Clinician's Guide to Evidence Base and Applications* (pp. 167–189). Burlington, MA: Academic Press.

Wurtzen, H., Dalton, S., Elsass, P., Sumbundu, A., Steding-Jensen, M., Karlsen, R., Andersen, K., Flyger, H., Pedersen, A., & Johansen, C. (2012). Mindfulness significantly reduces self-reported levels of anxiety and depression: results of a randomised controlled trial among 336 Danish women treated for stage I–III breast cancer. *European Journal of Cancer, 49:* 1365–1373.

Zainal, N., Booth, S., & Huppert, F. (2012). The efficacy of mindfulness-based stress reduction on mental health of breast cancer patients: a meta-analysis. *Psycho-oncology, 22:* 1457–1465.

Zeidan, F., Johnson, S., Diamond, B., David, Z., & Goolkasian, P. (2010). Mindfulness meditation improves cognition: evidence of brief mental training. *Consciousness and Cognition, 19:* 597–605.

INDEX